T3-BSK-123

Managing agrodiversity the traditional way

This book presents part of the findings of the international project "People, Land Management, and Environmental Change", which was initiated in 1992 by the United Nations University. From 1998 to 2002, the project was supported by the Global Environment Facility with the United Nations Environment Programme as Implementing Agency and the United Nations University as Executing Agency.

Managing agrodiversity the traditional way: Lessons from West Africa in sustainable use of biodiversity and related natural resources

Edited by Edwin A. Gyasi, Gordana Kranjac-Berisavljevic, Essie T. Blay, and William Oduro

TOKYO • NEW YORK • PARIS

United Nations University Press
The United Nations University, 53-70, Jingumae 5-chome,
Shibuya-ku, Tokyo, 150-8925, Japan
Tel: +81-3-3499-2811 Fax: +81-3-3406-7345
E-mail: sales@hq.unu.edu (general enquiries): press@hq.unu.edu
www.unu.edu

United Nations University Office at the United Nations, New York
2 United Nations Plaza, Room DC2-2062, New York, NY 10017, USA
Tel: +1-212-963-6387 Fax: +1-212-371-9454
E-mail: unuona@ony.unu.edu

United Nations University Press is the publishing division of the United Nations University.

Cover design by Rebecca S. Neimark, Twenty-Six Letters

Printed in the United States of America

UNUP-1098
ISBN 92-808-1098-7

Library of Congress Cataloging-in-Publication Data

Managing agrodiversity the traditional way : lessons from West Africa in sustainable use of biodiversity and related natural resources / edited by Edwin A. Gyasi . . . [et al.].
p. cm.
Includes index.
ISBN 92-808-1098-7 (pbk.)
1. Agrobiodiversity—Africa, West. 2. Agrobiodiversity—Africa, West—Case studies.
3. Biological diversity conservation—Africa, West. I. Gyasi, Edwin A. (Edwin Akonno), 1943–
S494.5.A43M352 2004
333.95′16′0966—dc22

2004014198

Contents

List of tables and illustrations

Tables

Figures

List of colour plates

The colour plates are indicated by (c) in the first instance in the text and are grouped together in the centre of the book. (Please refer to all plates in this manner, Plate 1(c), Plate 2(c) etc.)

Maps

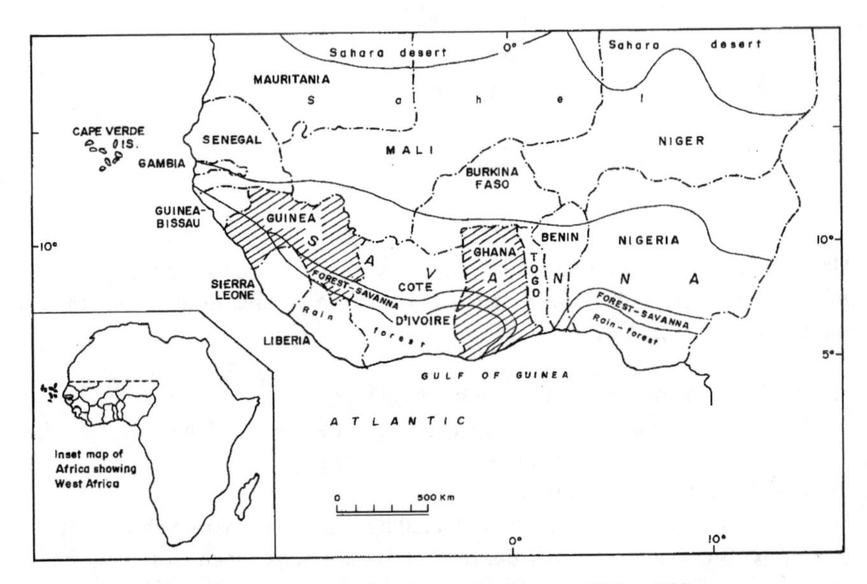

Map A Major ecological/vegetation zones of West Africa

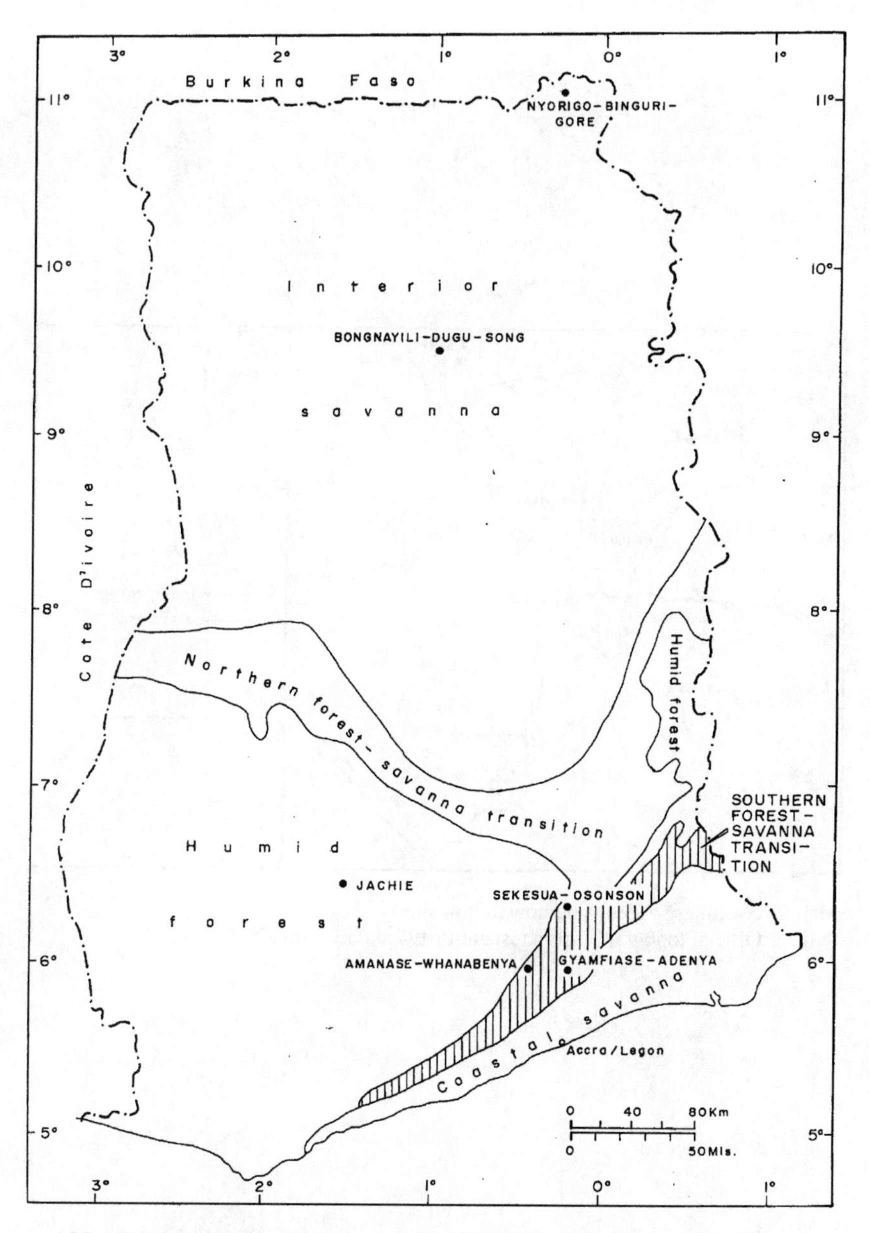

Map B Major agro-ecological zones and PLEC demonstration sites in Ghana

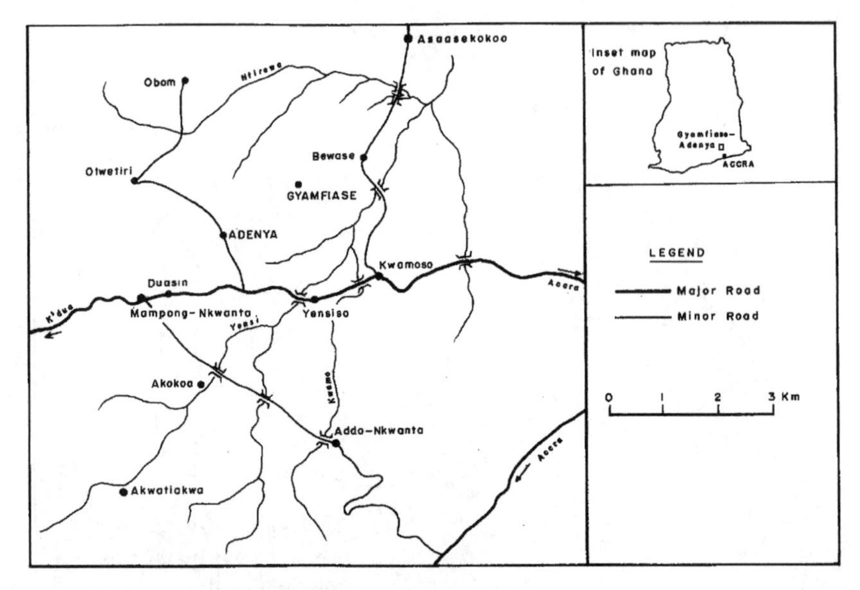

Map C Gyamfiase-Adenya demonstration site
Source: Official topographical maps and PLEC survey, 2000/2001

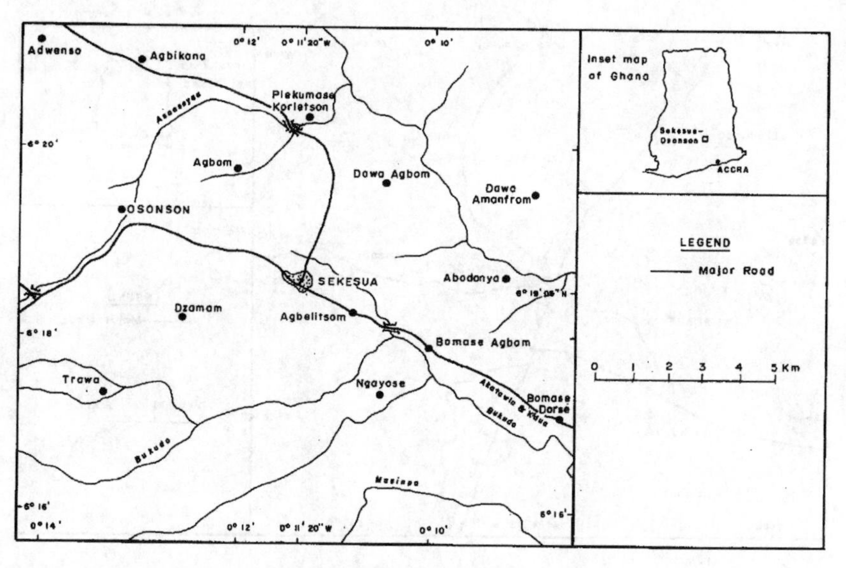

Map D Sekesua-Osonson demonstration site
Source: Official topographical maps and PLEC survey, 2000/2001

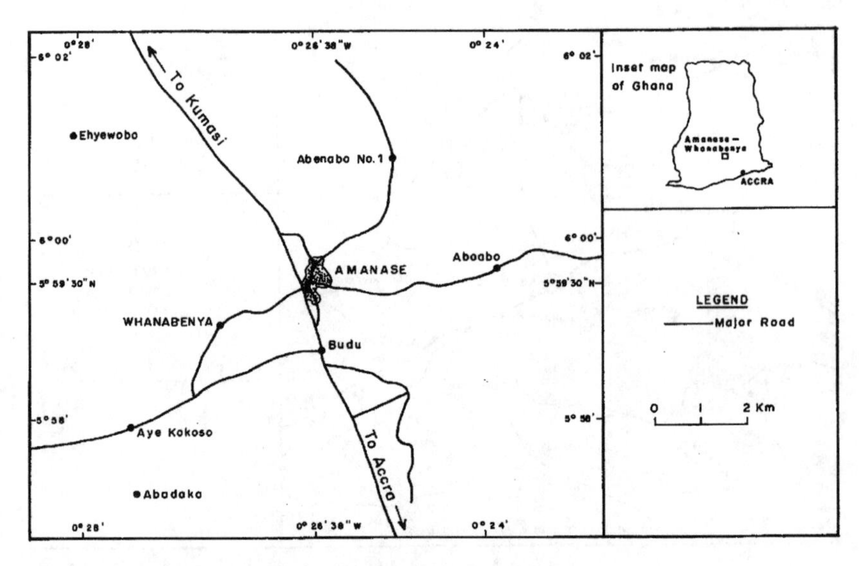

Map E Amanase-Whanabenya demonstration site
Source: Official topographical maps and PLEC survey, 2000/2001

Preface

This book is a product of work under the United Nations University Project on People, Land Management, and Environmental Change (modified to People, Land Management, and Ecosystem Conservation since 2002) – UNU/PLEC or PLEC, for short.

In 1994, barely a year after the inception of PLEC, its Scientific Advisory Group (SAG) wrote:

> human and social demands make it necessary that maintenance of biodiversity has to be accomplished within land use and agricultural systems in which farmers make use of a wide range of both natural and domesticated plant species. PLEC gives emphasis to the study of this agrodiversity, which is greatest among small-farming communities in the tropics and sub-tropics. Many farmers use indigenous knowledge and initiative as well as new information, to manage their land, waters and biota for production. Our [i.e. The PLEC] objective is to draw lessons as to which techniques and types of land use best perform the function of protecting natural resources, including the protection of a diversity of gene pools from which tomorrow's innovations may stem ...
>
> ... The ultimate [PLEC] aim is to provide researched options for the better management of land and resources for small-scale producers. Effective management systems do not have to be invented only by modern science. They exist, and have been continuously developed by the world's farmers. (Scientific Advisory Group, 1994)

The PLEC research across the tropical world confirms that inherent in used biophysical environments are indigenous, endogenous, local, or traditional practices that favour conservation of biodiversity through agrodiversity, i.e. agricultural diversification in all its forms – management diversity, agrodiversity, biophysical diversity, and organizational diversity (Brookfield, Stocking, and Brookfield 1999). However, this cultural heritage, which is fundamentally important for the ecological stability and genetic pool of plants and animals needed for food security, has come under threat mainly because official policy and modern development planners and practitioners alike largely ignore it. A lack of emphasis upon traditional knowledge in educational curricula associated with exotic values and the absence of relevant textbooks are contributory factors.

On the basis of case studies carried out under the nearly 10 years of PLEC multidisciplinary, participatory research work in three major agro-ecological zones (forest, savanna, and forest-savanna mosaic) in West Africa (mainly Ghana; Maps A and B), this book shows how, traditionally, farmers cultivate and conserve biodiversity while, at the same time, using the land for food production. It highlights PLEC interventions for sustaining agrodiversity for rural livelihoods, as it does lessons for teaching, policy, and development planning.

The book responds to various national as well as international policies and programmes on the environment, notably Ghana's Environmental Action Plan (Environmental Protection Council, undated), National Biodiversity Strategy for Ghana (Ministry of Environment and Science, 2002), and the Convention on Biodiversity (CBD; United Nations Environment Programme, 1992).

It would appeal to policy-makers and practitioners alike, and to university students and teachers, including those of agriculture, social science, biological science, and others relating to environmental or natural resources management and sustainable development.

In this book, the relevant research findings and their contextual background are presented in four parts, each containing a set of chapters.

Part I, comprising five chapters, focuses on "Methodological Approaches and Knowledge Systems". In the first chapter, Gyasi discusses the methodological approaches to the book in the context of the purpose and historical evolution of PLEC. Oteng-Yeboah follows up in Chapter 2 with an overview of philosophies that, traditionally, underlie the use of biophysical resources with special reference to Ghana. Then, in Chapter 3, Buabeng discusses traditional methods of resource assessment by farmers relative to the modern systematic methods by scientists. The discussion brings home the commonalities and dichotomies between the two. How farmers manage agrodiversity for human food security and survival in the wake of variable climatic conditions is still an ill-understood issue. Ofori-Sarpong and Asante address it in Chapter 4 with reference to PLEC demonstration sites in southern Ghana. The expert farmer and demonstration site notions, as discussed by Gyasi in Chapter 5 to conclude Part I, point to possibly cost-effective ways of recognizing, tapping, demonstrating, and upscaling conservation practices of exceptionally knowledgeable farmers.

Part II, titled "Cropping Systems and Related Case Studies", is made up of a set of 12 chapters that focus on system of managing crops and associated case studies. It opens with Chapter 6, where Gyasi profiles techniques and methods of managing biodiversity in farmed areas within PLEC demonstration sites in southern Ghana. It forms a useful background to the next two chapters, by Blay and by Kranjac-Berisavljevic and Gandaa, which provide insights into diverse ways of managing and conserving the diversity of yams in demonstration sites within the ecologically contrasting semi-humid forest-savanna mosaic zone of southern Ghana and semi-arid northern savanna zone. By focusing on the conservation of indigenous rice varieties by women in Chapter 9, Tanzubil, Dittoh, and Kranjac-Berisavljevic introduce a refreshing gender perspective. The chapter highlights the often unrecognized leading role of women in crop production and conservation of land-races.

In Ghana, rice and other starchy staples are commonly consumed together with sauces and soups prepared from vegetables. Managing the diversity of vegetables for food security with special reference to southern Ghana forms the theme of Chapter 10 by Blay.

There are, in Ghana, various traditional ecologically based smalholder farming systems for managing crops. Foremost among them is *proka* or *oprowka* (both expression from the Akan-Twi language) which conserves biodiversity and maintains soil fertility by using cleared vegetation for mulching instead of burning it off. In Chapter 11, Quansah and Oduro discuss the *proka* system with reference to Tano-Odumase, a PLEC demonstration site in central Ghana. Other ecologically based systems include home gardening, which, as discussed on the basis of experiences in PLEC demonstration sites in southern Ghana by Enu-Kwesi, Gyasi, and Vordzogbe in Chapter 12, has the advantage of both providing food security and serving as a germplasm bank. Similarly ecologically compatible is the practice of managing trees in association with crops in highly biodiverse traditional agroforestry systems, as discussed by Poku in Chapter 13.

A pertinent issue is the relationship between traditional farming practices and crop yields. In Chapter 14, Asafo *et al.* present preliminary investigations into this issue with reference to *proka* and to tree-crop combinations in Gyamfiase-Adenya in the semi-humid forest-savanna zone in southern Ghana. A similar issue is addressed by Anane-Sakyi and others in Chapter 15, but with a focus on effects of endemic trees on soil fertility in the drier northern savanna. Savanna woodlands cover extensive stretches of the middle portions of West Africa. In Chapter 16, Enu-Kwesi and Ghanaian and Guinean colleagues draw on information from comparative field work in Ghana and Guinea to show how, through various management systems, human beings are influencing the composition of the savanna woodland.

Without a strong incentive, farmers can hardly be expected to cultivate and conserve agrodiversity. In Chapter 17, Blay *et al.* discuss the management of forested and non-forested areas for honey production through apiculture, and for production of snails, the grasscutter, and yams so as to improve rural livehoods and incomes and, thereby, motivate agrodiversity.

Three chapters are brought together in Part III under the general rubric "Social Dimensions of Resource Management". The first one, Chapter 18 by Gyasi and

Asante, examines how biodiversity relates to resource tenure in both southern and northern Ghana. Chapter 19 by Bakang, Oduro, and Nkyi focuses on resources access in relation to the distribution and use of land in Tano-Odumase in central Ghana. Agbenyega and Oduro broaden the gender perspective in Chapter 20, by discussing the role of women in environmental management at Jachie, the premier PLEC demonstration site in central Ghana.

Finally, in Part IV, Gyasi, the leading editor, brings the book to a conclusion by pooling, from the preceding chapters, lessons for sustainable management of agrodiversity and related natural resources, and highlighting possible directions of future research work on them.

Should the findings presented in this book advance resource management knowledge beyond what is embodied in the maiden PLEC book (Gyasi and Uitto, 1997), its basic purpose would have been served.

Edwin A. Gyasi
Principal Editor

REFERENCES

Brookfield, H., M. Stocking, and M. Brookfield, "Guidelines on agrodiversity assessment in demonstration site areas (revised to form a companion paper to the BAG guidelines)" *PLEC News and Views*, Special Issue on Methodology, Vol. 13, 1999, pp. 17–31.

Environmental Protection Council, *Environmental Action Plan*. Accra (undated).

Gyasi, E. A. and J. I. Uitto, eds, *Environment, Biodiversity, and Agricultural Change in West Africa: Perspectives from Ghana*, Tokyo: United Nations University Press, 1997.

Ministry of Environment and Science, National Biodiversity Strategy for Ghana. Accra, 2002.

Scientific Advisory Group, "Population [subsequently *People*], Land Management and Environmental Change (PLEC) – A short statement by the Scientific Advisory Group", *PLEC News and Views* No.2, 1994, p. 1.

United Nations Environment Programme, *Convention on Biological Diversity*, Environmental Law and Institutions Programme Activity Centre Nairobi; United Nations Development Programme, 1992. See also, United Nations Environment Programme, *The Convention on Biological Diversity: Issues of Relevance to Africa*, Regional Ministerial Conference on the Convention on Biological Diversity. Nairobi: United Nations Environment Programme, 1994.

Foreword

Harold Brookfield

The PLEC group in Ghana was not only one of the project's first clusters to be formed; it was beyond dispute the first to produce substantial outputs. By the end of 1994, the pioneer group in Legon had completed an important pilot study in south-eastern Ghana and held an international workshop to which they brought some of those who became the project's first expert farmers. The product was brought together in the first book published from any part of PLEC (Gyasi and Uitto, 1997).

WAPLEC (West African cluster of PLEC) in Ghana has continued to be a major producer of research outputs, but most of those that have been published have appeared only in the unrefereed *PLEC News and Views*. A great deal more than these papers has reached my office, but mostly in the form of manuscript reports, not finalized in a publishable form; over the years, I have received more than 50 such reports. I was therefore delighted to learn, in 2000, that Edwin Gyasi was proposing to bring a major part of this work together in a new book.

This book, now that it is finished, is much more ambitious than its 1997 predecessor. It covers a large part of the work done by members of the large Ghana cluster over a 10-year period. Moreover, it puts this work into the larger context of international PLEC objectives and methods, policies of the government of Ghana, and international undertakings, specifically the Convention on Biological Diversity. It ranges through topics as varied as biodiversity inventory, agrobiodiversity conservation and promotion, the gender and land-tenure relations of production and management, specific management technologies that have been described and evaluated by the cluster's scientists, coping with climatic change, and the promotion of farmer initiatives that both help conserve diversity and obtain value from doing so. It describes, in some detail, work done in all seven of the principal and subsidiary demonstration sites that have been developed and sustained.

A high proportion of the book concerns these seven demonstration sites and the work done in and around them. PLEC in Ghana has employed a fairly specific definition of a demonstration site, and its methods have differed somewhat from those used elsewhere in the project. Chapters 5, 6, and 21 sum up both what has been learned about management methods and the numerous initiatives taken by PLEC in advancing development with conservation in the sites. The chapters on cropping systems, agrodiversity, and related case studies in Part II are the core of the book, preceded by general discussion of methodological approaches and knowledge systems in Part I, and followed up with a set of case studies on resource tenure and women's role under the general rubric "Social Dimensions of Resource Management" in Part III. There is an important group of chapters, fairly well distributed through the book, on the scientific work done to evaluate specific management practices, which has been a distinctive feature of the work of the Ghana cluster. These include the only chapter, which also draws on work in Guinea, reporting a comparative study of the diversity of savanna forest in relation to different systems of management.

There is particular emphasis on how much the practice of resource management for development can be informed by the innovations and adaptations evolved by the farmers themselves. Some of these practices have been under threat of loss through disuse in recent years, and it has been a major thrust of PLEC work in the last three years to give them new vigour, while validating their scientific value. The outstanding case is the no-burn *oprowka* or *proka* system of land preparation, introduced in Chapter 6 and the topic of two scientific studies in Chapters 11 and 14. Readers will note that the findings of these two studies are somewhat at odds, indicating the amount of work that is still to be done. The Ghanaian farmers, both men and women, are presented in these pages as skilled, astute, and innovative in the face of quite difficult conditions of uncertain rainfall, low and sometimes declining soil fertility, rising pressures of population on resources, and the effects of globalization on the market for their crops.

It has been my good fortune to visit Ghana five times during the PLEC years, usually to attend one or other of the annual workshops of the cluster. I have been able to visit all but one of the demonstration sites and to witness their progress as evolving community-based organizations for conservation and development. I also twice went to Guinea, and regret that the work done there could not be adequately represented in this book, as was originally intended. Some Ghanaian chapters that were originally planned are also missing from the final assembly. But the book had to be finished within a limited time, and the editors are to be congratulated on their achievement in bringing together so much of the work done by PLEC's largest and perhaps most vigorous cluster. The principal editor writes that "should the findings presented in this book advance resource management knowledge beyond what is embodied in the maiden PLEC book [of 1997], its basic purpose would have been served". It does this abundantly, presenting also a fine example of the achievement of Ghanaian scientists working together with Ghanaian farmers.

REFERENCES

Gyasi, E. A. and J. I. Uitto, eds, *Environment, Biodiversity and Agricultural Change in West Africa: Perspectives from Ghana*, Tokyo: United Nations University Press, 1997.

Acknowledgements

Edwin A. Gyasi, Gordana Kranjac-Berisavljevic, Essie T. Blay, and William Oduro

This book is a product of collaborative research work by scientists, farmers, and policy agents under the United Nations University project on People, Land Management, and Environmental Change (UNU/PLEC), since the year 2002 redesignated People, Land Management, and Ecosystem Conservation.

Mainly work was carried out over the four-year period from 1998/99 to 2001/02, with funding by the Global Environment Facility (GEF) through the UNU and the United Nations Environment Programme (UNEP). Without this support, the book probably would never have seen the light of the day. Therefore, we wish to register our profound gratitude to the GEF for its financial support, and to the UNU and UNEP for facilitating that support.

Since the inception of PLEC in West Africa, the University of Ghana has served as the management centre for work in southern Ghana and as the principal administrative node for the country. In the initial phase, the administrative responsibility covered the whole of West Africa. For discharging this role efficiently, we thank the University, particularly its Consultancy Centre for managing project funds and the Department of Geography and Resource Development for housing the project.

In central Ghana PLEC work is managed through the Institute of Renewable Natural Resources (IRNR), Kwame Nkrumah University of Science and Technology (KNUST), Kumasi; in northern Ghana through the University for Development Studies (UDS), Tamale; and in the Republic of Guinea through the Centre d'Etude et de Recherché en Environnement (CERE), Universite de Conakry. We are deeply appreciative of their facilitating role.

Staff of the UNU, most especially Liang Luohui, Ebisawa Masako, and Ichikawa Wakako, were very supportive, particularly during the critical final stages of the preparation of the manuscript. We thank them for their patience and gentle way of prodding us on.

For inspiring the entire UNU/PLEC and us intellectually, we wish to place on record our profound gratitude to Emeritus Professor Harold Brookfield, the principal PLEC scientific coordinator, of the Australian National University.

For motivating us by their mature advice, we are grateful to Ebenezer Laing, Professor Emeritus, Department of Botany, University of Ghana; George Benneh, Professor Emeritus, Department of Geography and Resource Development, University of Ghana; and Uzo Mokwunye, Professor and Director, United Nations University for Natural Resources in Africa (UNU/INRA).

Felix Asante, Kenneth Peprah, and Emmanuel Joseph Mensah, all from the University of Ghana, Legon, performed superbly as research assistants. We thank Felix most sincerely for helping to generate and analyse information, and Ken and Emma for their competent typing work.

The quality of the maps attest to the cartographical skills of D. J. Drah of the Department of Geography and Resource Development, University of Ghana, Legon. We thank him for his skill and willingness to deliver at short notice.

Each of the 21 chapters makes a unique contribution to the book. To their authors we say a big thank you for their effort, as we do for the constructive comments of the two anonymous reviewers of the book manuscript.

But, above all, it is the hundreds of PLEC farmers whose knowledge forms the fundamental source of information. We are profoundly grateful for their knowledge of resource management and for their hospitality during our numerous visits. As ever, these unsung farmers were always prepared to share. To them we dedicate this book as a token recognition.

Part I

Methodological approaches and knowledge systems

1

Methodological approaches to the book

Edwin A. Gyasi

Agrodiversity defined

Through case studies in West Africa (principally Ghana; Maps A and B) and drawing from nearly 10 years of research experience of the United Nations University project on People, Land Management, and Environmental Change (UNU/PLEC), this book demonstrates the importance of traditional, indigenous, or local farmer knowledge and practices in sustainable conservation of biodiversity and related natural resources by agrodiversity.

Agrodiversity refers to the processes and products of agricultural diversification. In more elaborate terms, it is "the many ways in which farmers use the natural diversity of the environment for production, including not only their choice of crops but also their management of land, water and biota as a whole" (Brookfield and Padoch, 1994: 9). It comprises four principal elements, namely:

- management diversity, which refers to the various methods of managing the land and associated biophysical resources for agricultural purposes
- agrobiodiversity, which describes the "management and direct use of biological species, including all crops, semi-domesticates and wild species" (Huijun, Zhiling, and Brookfield, 1996: 15)
- biophysical diversity, which refers to the various soil characteristics and their productivity, and the biodiversity of natural (or spontaneous) plant life and the soil biota

- organizational diversity, which describes the diverse socio-economic aspects of farming such as tenurial arrangements, household characteristics, and gender roles (Brookfield, Stocking, and Brookfield, 1999; see also Brookfield 2001; Brookfield *et al.*, 2002).

In varying degrees, the case studies that form the core of the book illustrate all four elements of agrodiversity.

It is believed that agrodiversity has the virtue of:

- strengthening biodiversity
- imparting ecological stability
- providing a genetic pool of plants and animals needed for breeding higher-yielding varieties for food security
- facilitating dietary diversity
- contributing to "(1) increased resource productivity over time, (2) increase in the amount and quality of labour applied to the farm, and (3) insurance and risk reduction at household enterprise level" (Netting and Stone, 1996: 53; see also the other articles in *Africa*, Vol. 66, and Brush, 2000).

Historical context

The mainly econcentric case studies of the book (Jones and Hollier, 1997) are rooted in the pilot West African PLEC (WAPLEC) work, which was initiated in the year 1993, with a focus on understanding agro-environmental changes and farmers' role in and reactions to them.

That initial, basically investigative work involved studies in three principal sites, centred on Yensiso (subsequently renamed Gyamfiase-Adenya), Sekesua (subsequently renamed Sekesua-Osonson), and Amanase (renamed Amanase-Whanabenya), all located in the southern sector of Ghana's forest-savanna transition zone (Maps B, C, D, E). The principal outputs of those studies, carried out by scientists from the University of Ghana, Legon, were:

- the development of research links with farmers and increased insights into agro-ecological transformations, which served as a basis for further work
- the scientific paper "Production pressure and environmental change in the forest-savanna of southern Ghana" (Gyasi *et al.*, 1995)
- the book *Environment, Biodiversity and Agricultural Change in West Africa: Perspectives from Ghana* (Gyasi and Uitto, 1997).

Subsequently, the work was extended to additional sites in Ghana's remaining major agro-ecological zones, namely humid forest and dry savanna, and in wooded savanna portions of the Fouta Djallon mountains of the Republic of Guinea (Maps A, B, C, D, E). The extension was facilitated by:

- integration of more scientists from the University of Ghana, and of additional ones from the Kwame Nkrumah University of Science and Technology

(KNUST) and University for Development Studies (UDS), both in Ghana, and the Université de Conakry in the Republic of Guinea
- increased collaboration with farmers.

From about 1997, the focus shifted to:
- identification of those aspects of farmer resource usage that appear to favour agrodiversity
- demonstration and improvement of sustainable agrodiversity management practices as a way of meeting simultaneously the triple objectives of conserving biodiversity, strengthening food security, and enhancing rural livelihoods.

Methodology

The ensuing case studies mainly contain the findings of the post-1997 PLEC research. In carrying out that work and the maiden work that preceded it, participatory procedures were commonly followed by the research scientists.

The participatory procedures involved learning farmer practices and their underpinning knowledge by close collaborative work between the multidisciplinary teams of scientists and the farmers through:
- group discussions
- farm visits
- joint on-farm experiments and other forms of cooperative ventures within the selected project focal sites (Map B).

These activities were facilitated by collaboration with governmental and non-governmental agencies, and by farmer associations in which, as discussed in Chapter 5, expert farmers played a central role, especially as sources of local knowledge and as mediators with other farmers. Overall, the farmer associations were composed of a mix of males and females numbering more than 1,300 people.

In the work with the farmers (PLEC members as well as non-PLEC members), special emphasis was placed upon understanding how, on the basis of traditional knowledge, farmers manage agrodiversity. Because traditional knowledge reflects local conditions including popular values, it can be assumed to offer a sounder basis for developing more locally adaptive resource management models in line with the grassroots, bottom-up development paradigm. Seen in this vein, traditional or indigenous knowledge may be said to be "complementary to conventional science" (Brokensha, Warren, and Werner, 1980: 8; see also Richards, 1985; Chambers, Pacey, and Thrupp, 1989; Benedict and Christofferson, 1996; Chambers, 1998; Mammo, 1999; Van den Breemer, Drijver, and Venema, 1995; Haverkort, van't Hooft, and Hiemstra, 2003).

All the three principal teams of PLEC research scientists based, respectively at the University of Ghana, Kwame Nkrumah University of Science and Technology, and the University for Development Studies equally followed the multidisciplinary approach. It involved discussions with farmers, biodiversity assessment (Zarin, Huijun, and Enu-Kwesi, 1999), computer programming of information generated, and the discharge of other work aspects on a joint basis by the teams of scientists and other experts drawn from a diversity of specializations – botany, soils science, crop science, other biophysical sciences, agricultural economics, geography, and other social sciences. But the team approach was pursued without sacrificing individual disciplinary perspectives, particularly with regard to the interpretation of the information generated through the multidisciplinary teamwork.

REFERENCES

Benedict, F. and L. E. Christofferson, eds, *Environment and Development in Africa: Participatory Processes and New Partnerships*, Copenhagen: Scandinavian Seminar College, 1996.

Brokensha, D. W., D. W. Warren, and O. Werner, eds, *Indigenous Knowledge Systems and Development*, Boston: University Press of America, 1980.

Brookfield, H., *Exploring Agrodiversity*, New York: Columbia University Press, 2001.

Brookfield, H. and C. Padoch, "Appreciating agrodiversity: A look at the dynamics and diversity of indigenous farming systems", *Environment*, Vol. 36, No. 5, 1994, pp. 6–11, 36–45.

Brookfield, H., M. Stocking, and M. Brookfield, "Guidelines on agrodiversity assessment in demonstration site areas (Revised to form a companion paper to the BAG guidelines)", *PLEC News and Views*, No. 13, 1999, pp. 17–31.

Brookfield, H., C. Padoch, H. Parsons, and M. Stocking, "Cultivating Biodiversity: Setting the scene" in H. Brookfield, C. Padoch, H. Parsons, and M. Stocking, eds, *Cultivating Biodiversity: Understanding, Analysing and Using Agricultural Diversity*, London: ITDG Publishing, 2002, pp. 1–8.

Brush, S. B., *Genes in the Field: On-Farm Conservation of Crop Diversity*, Boca Raton: Lewis Publishers, 2000.

Chambers, R., "Behaviour and attitudes: A missing link in agricultural science", in V. L. Chopra, R. B. Singh, and A. Varma, eds, *Crop Productivity and Sustainability, Proceedings of 2nd International Crop Science Congress*, New Delhi: Oxford University Press and IBH Publishing, 1998.

Chambers, R., A. Pacey, and L. A. Thrupp, eds, *Farmer First: Farmer Innovation and Agricultural Research*, London: Intermediate Technology Publications, 1989.

Gyasi, E. A. and J. I. Uitto, eds, *Environment, Biodiversity and Agricultural Change in West Africa: Perspectives from Ghana*, Tokyo: United Nations University Press, 1997.

Gyasi, E. A., G. T. Agyepong, E. Ardayfio Schandorf, L. Enu-Kwesi, J. S. Nabila, and E. Owusu-Bennoah, "Production pressure and environmental change in the forest-savanna zone of southern Ghana", *Global Environmental Change*, Vol. 5, No. 4, 1995, pp. 355–366.

Haverkort, B., K. van't Hooft, and W. Hiemstra, eds, *Ancient Roots, New Shoots: Endogenous Development in Practice*, Leusden: ETC/Compas, 2003.

Huijun, G., D. Zhiling, and H. Brookfield, "Agrodiversity and biodiversity on the ground and among the people: Methodology from Yunnan", *PLEC News and Views*, No. 6, 1996, p. 15.

Jones, G. and G. Hollier, *Resources, Society and Environmental Management*, London: Chapman Publishing, 1997.

Mammo, T., *The Paradox of Africa's Poverty: The Role of Indigenous Knowledge, Traditional Practices and Local Institutions – The Case of Ethiopia*, Lawrenceville and Asmara: Red Sea Press, 1999.

Netting, R. M. and M. P. Stone, "Agro-diversity on a farming frontier: Kofyar smallholders on the Benue plains of central Nigeria", *Africa*, Vol. 66, No. 1, 1996, pp. 52–70.

Richards, P., *Indigenous Agricultural Revolution: Ecology and Food Production in West Africa*, London: Hutchinson, 1985.

Van den Breemer, J. P. M., C. A. Drijver, and L. B. Venema, eds, *Local Resource Management in Africa*, Chichester: John Wiley & Sons, 1995.

Zarin, D. J., G. Huijun, and L. Enu-Kwesi, "Methods for the assessment of plant species diversity in complex agricultural landscapes: Guidelines for data collection and analysis from the PLEC Biodiversity Advisory Group (PLEC-BAG)", *PLEC News and Views*, No. 13, 1999, pp. 3–16.

2

Philosophical foundations of biophysical resource use with special reference to Ghana

Alfred A. Oteng-Yeboah

Traditional religious philosophical thoughts on use of the biophysical environment

Traditional religious philosophy considers the belief, doctrinal, spiritual, and/or worship systems as tools in ensuring the harmony of life even before the issues of the biophysical resources conservation and sustainable use come into play. Folk stories, local drama, and other local forms of communication are replete with information on how this is ensured.

Among the Akan people of Ghana, the philosophy has its foundations in the traditional concept of land ownership. Danquah (1968) indicates that the living, referring to present-day people, have an obligation to their ancestors to ensure a proper stewardship of the land for the use of future generations.

Benneh (1990) makes reference to an expression of sustainability of land made by *Nana* Sir Ofori Atta, in which the late chief conceives of land as belonging to a vast family of whom many are dead, a few are living, and countless hosts are still unborn. According to Abayie-Boaten (1999) this concept has shaped the perception of the African, in that through his relationship with nature, which is clothed in religion with its attendant reverent attitudes towards it, he has developed an attitude of a caretaker of his environment.

Significantly, the African regards the earth as a mother. In the Akan language, the earth is often addressed as *Asaase Yaa,* i.e. Mother Earth. Mother Earth is revered as the provider and sustainer of life. This is the philosophy behind the offer of prayers in the form of libation-pouring before cultivation of the land and during harvesting. Very often there are festivals, such as the Homowo (Ga people), Akwambo (Gomoa), Ohum (Akyem Abuakwa), Odwira (Akuapem), and Kundum (Nzema), at harvesting to commemorate the generosity of Mother Earth.

In order to allow Mother Earth continuously to play her mothership role and enhance the biodiversity that the earth contains, humanity has developed a body of laws and rules known as taboos to regulate its relationship with the environment (Abayie-Boaten, 1999).

Hagan (1999) lists the following seven traditional laws for the sustainable use of biosystems:

- laws of exclusion prohibiting entry into forests, lakes, and rivers except at periods of severe scarcity and critical needs
- laws of selective extraction protecting certain species or prohibiting the destruction and use of immature animals, e.g. pregnant animals were generally not killed for consumption
- laws governing diversification of use, to avoid over-exploitation of one or two crops or animals in the clan/community diet
- laws regulating exploitation, enforced by rites of closing and opening of rivers, lakes, estuaries, and forests under constant use, to enable the regeneration of species in ecosystems
- laws enforcing community involvement in land preparation for farming, to ensure the containment of fire hazards
- laws protecting special species of plants and animals from misuse, to ensure high stock levels, e.g. certain tree species are not to be felled for fuelwood
- laws enforcing rites for the felling of big trees and killing of certain animals, to ensure the protection of these organisms and also make the ecosystem secure.

In addition, environmentally low-impact systems and tools have been developed over the years for tillage of the land in a sustainable manner.

A notable law or taboo is the one that prohibits the felling of certain species of trees, e.g. "*odum*" – *Milicia excelsa*, and "mahogany" – *Khaya ivorensis*, unless some rituals are performed, because those trees are considered sacred. Busia (1954) recounts a typical story of tree ritual as follows:

> An Ashanti craftsman will endeavour to propitiate certain trees before he cuts them down. He will offer an egg, for example, to the *odum* tree, saying "I am about to cut you down and carve you; do not let me suffer any harm."

Similar rituals are performed on several other trees, and this helped to preserve several economic trees as well as other trees which were not economically useful but useful in maintenance of the environment. Such trees as "*onyina*", *Ceiba pentandra*, "*akonkodie*", *Bombax buonopozense*, and tall "*beten*", *Elaeis guinensis*, can be cited as examples of trees that enjoyed the kind of protection.

Other taboos or laws of prohibition include the following:

- certain forests are not to be entered into on certain days because of beliefs that the spirit being in those forests is roaming around at the time
- some animals are considered sacred or totem organisms and they are not to be hunted (Adarkwa-Dadzie, 1999; Telly, 1999; Voado, 1999)
- particular types of farm implements are not to be used because they impact negatively on the soil and tear roots
- snails are not to be picked at night because, being nocturnal, a whole population of snails may be out to feed and could be decimated
- certain types of food crops are not to be harvested and eaten until the performance of some rituals by the whole community
- in some communities, it is forbidden to farm and fell trees on slopes and in watershed areas.

The earth is expected to rest or get restored after use in one of the seven days of the week. During this time no one is allowed to till the land, or fish from the river. This practice is similar to the Sabbath observation in Christian and Jewish religions.

Each locality had its own day of rest. In the coastal areas ethnic groups had taboo days during which there should be no fishing. There is also a long period during which no fishing is expected to be done in lagoons and other coastal water bodies. This resting period coincides with the period when the fish in the lagoons lay their eggs.

Apart from the apparent fear of spirits, which these taboos instilled in the people, there were also physical sanctions against breaking the taboos. To ensure the propitiation of the spirit, which involved the spilling of blood, the culprit had to offer sheep and some bottles of schnapps, a liquor. These sanctions were considered deterrent to scare people from breaking taboos deliberately.

There was a taboo against the clearing of the vegetation right up to the edges of streams and rivers. The people were aware that this could check excessive evaporation from the rivers and streams.

Bush fallow farming, shifting cultivation, rotationary agroforestry, compound farming, home gardens, and other low-impact farming systems were encouraged. Gyasi (1999) considers these farming systems as mimicking the natural forest ecosystem, inherently self-regenerative, and/or protective of the soils and biological diversity because of their close adaptation to the local ecological niches and the natural biophysical environment they are designed to imitate.

Benneh (1997), while extolling the virtues of shifting cultivation as a good example of a traditional organic farming system, indicated that the system enjoys the advantages of:

- minimizing soil erosion
- preserving agrobiodiversity
- maintaining ecological stability

- optimizing utilization of the different soil nutrients
- enhancing food security and a balanced diet.

Gyasi (1991, 1994, 1999) has provided the following list of traditional farming practices that enhance ecological values and sound biotic and abiotic conservation:

- intercropping and rotations incorporating useful medicinal and other natural plants that are conserved *in situ* as part of the cropping systems
- integration of nitrogen-fixing herbaceous and leguminous crops
- nurturing of certain useful naturally propagated trees
- nurturing of useful plants by both *in situ* and *ex situ* methods in home gardens and outfields
- prohibition of the felling of certain tree species for their spiritual, economic, and ecological values
- use of mounds for soil-moisture conservation
- teaching conservation to children
- preservation of trees and other plants whose presence is indicative of good soil, which enhance soil moisture, or which provide ideal shading conditions for shade-loving crops.

The thoughts behind these traditional or religious philosophies, though in some instances shrouded in secrecy and myths to frighten people, were well intended. They were meant to ensure a sustainable harmony of human beings with their environment for the purpose of survival.

Though the reasons behind many of these thoughts were never explained traditionally, they are now being explained scientifically. Many of these are now being rediscovered to be used in conservation programmes.

Modern scientific philosophical thoughts on the use of the biophysical resource

Modern scientific thought considers, as viable, a system of the environment in which the net flow of energy and nutrients is in dynamic equilibrium with the functions of the components of the biophysical environment. In other words, ecosystem integrity, resulting from a healthy balance between the various physical, chemical, and biological cycles, is responsible for the contribution to life sustenance of the biophysical environment.

The understanding of this philosophy is that there must be a direct relationship between the physical cycles (including the chemical and biological cycles), which must provide the basis for the conservation and sustainable use of the biophysical resources. This determines how the exploitation of the biological resources and physical resources can be conducted to ensure sustainability.

Conclusion

What is being established in all cases of biophysical resources' exploitation is the need for sustainability. The concept of sustainable biophysical resource use therefore implies the use of practices (traditional religious and/or scientific) that have success stories attached to them, that are being applied in conservation and sustainable use programmes internationally and locally, and that can be empirically verified and certified.

The works of the West African cluster of UNU/PLEC provide specific examples in which the biophysical environment and farmers' responses to them in West Africa, with special reference to Ghana, have been captured. The expectations are that these interesting case studies will be carefully digested and their recommendations implemented to ensure the sanctity of the environment for sustainable rural livelihoods and food security.

REFERENCES

Abayie-Boaten, N. A., "Traditional conservation practices: Ghana's example", in D. S. Amlalo, L. D. Atsiatorme, and C. Fiati, eds, *Biodiversity Conservation: Traditional Knowledge and Modern Concepts*, Accra: Environmental Protection Agency (EPA), 1999, pp. 1–6.

Adarkwa-Dadzie, A., "The Contribution of Ghanaian traditional beliefs to biodiversity conservation", in D. S. Amlalo, C. D. Atsiatorme, and C. Fiati, eds, *Biodiversity Conservation: Traditional Knowledge and Modern Concepts*, Accra: Environmental Protection Agency (EPA), 1999, pp. 30–32.

Benneh, G., "Towards sustainable development: An African perspective", *Geografish Tidsslerift*, Vol. 90, 1990, pp. 1–4.

Benneh, G., "Indigenous African farming systems: Their significance for sustainable environmental use", in E. A. Gyasi and J. I. Uitto, eds, *Environment, Biodiversity, and Agricultural Change in West Africa*, Tokyo: United Nations University Press, 1997, pp. 13–18.

Busia, K. A., "The Ashanti of the Gold Coast in African world studies", in D. Forde, ed., *The Cosmological Ideas and the Social Values of African Peoples*, London: Oxford University Press, 1954.

Danquah, J. B., *The Akan Doctrine of God*, London: Frank Cass and Co., 1968.

Gyasi, E. A., "Communal land tenure and the spread of agro-forestry in Ghana's Mampong valley", *Ecology and Farming*, Vol. 2, 1991, pp. 16–17.

Gyasi, E. A., "The adaptability of African communal land tenure to economic opportunity: The example of land acquisition for oil palm farming in Ghana", *Africa*, Vol. 64, No. 3, 1994, pp. 391–405.

Gyasi, E. A., "Land tenure system and traditional concepts of biodiversity conservation", in D. S. Amlalo, L. D. Atsiatorme, and C. Fiati, eds, *Biodiversity Conservation: Traditional Knowledge and Modern Concepts*, Accra: Environmental Protection Agency (EPA), 1999, pp. 16–23.

Hagan, G. P., "Traditional laws and methods of conservation and sustainable use of biodiversity", in D. S. Amlalo, L. D. Atsiatorme, and C. Fiati, eds, *Biodiversity Conservation: Traditional Knowledge and Modern Concepts*, Accra: Environmental Protection Agency (EPA), 1999, pp. 24–29.

Telly, E. M., "Some traditional mechanisms for protecting the environment", in D. S. Amlalo, L. D. Atsiatorme, and C. Fiati, eds, *Biodiversity Conservation: Traditional Knowledge and Modern Concepts*, Accra: Environmental Protection Agency (EPA), 1999, pp. 48–54.

Voado, G. C., "Some perspectives of traditional African knowledge in biodiversity conservation", in D. S. Amlalo, L. D. Atsiatorme, and C. Fiati, eds, *Biodiversity Conservation: Traditional Knowledge and Modern Concepts*, Accra: Environmental Protection Agency (EPA), 1999, pp. 43–54.

3

Traditional methods of resource assessment relative to the scientific approach

Stephen Nkansa Buabeng

Introduction

This chapter examines traditional methods of resource assessment relative to the scientific with reference to Jachie and Tano-Odumasi, two PLEC demonstration sites in central Ghana (Map B). The main objective is to find out how the indigenous method of assessment differs from the scientific approach, and how the two approaches may be bridged to enhance the capacity of the farmer in improving natural resource management.

The contribution of traditional indigenous knowledge to development was very much ignored in the past. But, in recent times, it has been recognized that it is very essential in the development process. This is evidenced in the work of PLEC in various areas in Ghana, and elsewhere in the world.

The concept of participatory development, which ensures traditional people's expression of ideas or project beneficiaries' input into project design, is a welcome counter to attempts at the marginalization of traditional knowledge.

Relevant data for the chapter were generated by participatory rural appraisal methods. They include focus group discussions with PLEC farmers combined with field observations.

Traditional knowledge defined

Traditional or indigenous knowledge is the knowledge which has been acquired over the years through experience, and passed on from generation to generation through oral tradition and by practice. It can be divided into two, namely technical and non-technical. Technical knowledge refers to that knowledge which incorporates skills and is manifested in production systems and socio-cultural systems such as arts and music. Non-technical knowledge refers to value systems, beliefs, customs, and rules of behaviour. Hountondji (1997) has described traditional technical knowledge research as ethno-technology, and within the realm of science as ethno-science, which is defined as the study of a corpus of knowledge, information, and know-how handed down from generation to generation. Examples of these are ethno-zoology, the study of traditional zoological knowledge, ethnobotany, the study of traditional botanical concepts, and ethno-minerals, the study of traditional knowledge of mineral resources. Following this concept, PLEC activities can be described as ethno-biodiversity, the study of traditional knowledge in biodiversity conservation and management systems.

Traditional knowledge has several characteristics, including the following:

- its qualitative nature
- its location specificity, i.e. traditional knowledge relates naturally to a specific place
- its coverage of several areas of human activities, such as cultural, social, and production organizations, festivities, music, religion, land management, and conservation.

Indigenous knowledge of any group of people, especially rural people, is made accessible to outsiders only through learning from such people themselves or, occasionally, through ethnographic literature using anthropological jargon. Furthermore, indigenous knowledge exists in immeasurable forms among immeasurable groups of people in immeasurable environments.

Scientific knowledge, on the other hand, refers to knowledge that is based on, regulated by, or done according to facts and the laws of science. Some important characteristics of scientific knowledge are that it is systematic and exact. It is obtainable or accessible to other people or can be shared with other people through books, journals, and information-retrieval systems. Scientific knowledge is therefore communicated, taught, and is available all over the world and, therefore, has universal acclaim.

The development of information technology, specifically the internet, has made both scientific and indigenous knowledge accessible to a wider audience.

Scientific approach to measurement of soil fertility

There are several scientific ways of assessing soil fertility and classifying soils. The classification is often based on the use to which the soil is going be put. It may be based on the physical, chemical, or biological properties of the soil. The measurements are often based on routine laboratory techniques.

Physical measurements

In order to characterize soils by their physical properties the following indicators may be routinely monitored:

- texture
- bulk density
- field capacity
- available water capacity
- infiltration.

Soil texture is commonly determined because it affects crop performance Texture or particle size distribution refers to the relative proportion of sand, silt, and clay in soil. In the laboratory these fractions are determined by sieving and sedimentation methods. Depending on the relative proportions of sand, silt, and clay, soil may be classified as shown in Table 3.1. In the field, experienced soil scientists and technicians can use the "feel method", rubbing some soil between the thumb and the forefinger, to determine the texture of the soil.

Soil chemical and biological measurements

Together with the soil physical measurements, the following parameters may be determined by routine chemical analyses to determine the level of soil fertility.

- pH
- exchanging cations

Table 3.1 Scientific classification of soils

S. no.	Common names	Texture	Basic soil textural class names
1	Sandy soils	Coarse	Sandy Loamy sands
2	Loamy soils	Moderately coarse	Sandy loam Fine sandy loam
		Medium	Very fine sandy loam Loam Silt loam Silt
		Moderately fine	Clay loam Sandy clay loam Silty clay loam
3	Clayey soils	Fine	Sandy clay Silty clay Clay

- effective cation exchange capacity (ECEC)
- organic carbon
- nitrogen
- phosphorus
- potassium.

The levels of these parameters are compared with some standards to indicate whether the soil is fertile or not. The standards vary with the method used in the laboratory analyses. A soil which is high in the above parameters is considered fertile. In order to measure these parameters, soil samples are taken from the field and conveyed to the laboratory for the analysis. It must be emphasized that the quality of the results is dependent on the quality of laboratory analysis.

Indirect methods

In addition to the scientific method, which requires routine laboratory analysis, scientists employ indirect methods in assessing soil fertility by the use of certain parameters such as colour. A soil that is black or dark in colour may indicate a high amount of organic matter or humus and high fertility status. A red soil indicates that the soil is well drained, well aerated, and contains ferric oxide. Such soils are often seen at summits and upper to middle-slope sites. Brown to light brown soils, often found in the middle to lower-slope sites, are moderately well drained. Grey or white soils, often found between the lower slopes and the valley bottoms, are poorly drained and subject to waterlogging.

Some characteristics of sand and clay soils

Sandy soils contain a large number of big pores. Therefore they are well aerated. Drainage is very good but they are poor in retaining moisture and nutrients. They are generally considered to be of poor fertility. They require addition of organic manure and/or mineral fertilizers to make them fertile. They can easily be tilled with machines. Clay soils are poorly drained and aerated. They have a high capacity to store water and nutrients, but it is not easy for plants to abstract water from clay soils.

Community approach to measurement of soil fertility and classification

In assessing a natural resource such as soil fertility, a community's main method is observation. The observation is based on certain parameters such as colour, indicator plants, micro fauna and wildlife, and land-use stage.

Physical characteristics

In direct observation, farmers use the physical elements of colour, surface characteristics, and texture of the soil in determining soil fertility. The colour of the soil is used as a basis for clarification. The colours involved are white, black, and brown or red. Besides direct observation the farmers use the feel method by rubbing the sand in their palms.

The farmers have a basic knowledge about which types of crops are suitable for a particular soil. This knowledge is acquired through observation and information passed on from oral tradition. For example, the *afowiaah* (local name), which is whitish, is suitable for vegetables, cassava, maize, and plantain, and the gravelly soils are suitable for cassava. The valley bottoms soils, locally known as *wora*, are good for dry-season vegetables.

Indicator plants

The use of indicator plants is very common among farmers in determining soil fertility. For instance, areas dominated by *Chromolaena odorata* (especially where there is no undergrowth) and elephant grass (where they are bunchy) have good soil fertility. On the other hand the presence of spear grass (*Imperata* and *Paniacum rotboelia*) indicates poor fertility. The farmers are also aware of certain tree species, the presence of which improves the soil fertility and crop yields, e.g. *onyina – Ceiba pentandra*. These types of trees are left standing on the farm during land clearing. However, because of land degradation some of these cannot be found in the demonstration sites (Buabeng, 1998). Table 3.2 shows some of the indicator plants.

Micro-fauna, wildlife, and vegetative cover

The third method of assessment is the use of micro-fauna. The presence of worm casts and worm borings indicates that the soil is fertile. The presence of wildlife, both arboreal and terrestrial animals, indicates good soil fertility. The fourth method is the use of vegetative cover. There is no doubt about primary forest soil fertility.

Table 3.2 Indicator plants

Trees	
Local name	Biological name
Yankyere	*Ficus varifolia*
Funtum	*Funtumia elastica*
Onyina	*Ceiba pentandra*
Duma	*Ficus capensis*
Grasses	
Acheampong	*Chromolaena odorata*
Other grasses	*Imperata*
Herbs	
Nsansono (itching plant)	
Ananse Ntromuhoma	

Farmers use fallow periods to restore the fertility of the soil. The early period of fallow, which is called *forbe* regrowth, is a good indicator of soil fertility. Another vegetative parameter used is the presence of twigs/shrubs and creepers. High amounts of these indicate good soil fertility.

Comparison of traditional and scientific methods of assessment

Water quality

Traditional assessment of water resources is dependent on a number of factors. They include flow rate, the presence or absence of sand in the water or stream, and absence of vegetative cover trees and shrubs around it. Regarding water quality, parameters used include the colour of the water, the amount of foreign matter in the water, and the taste. By contrast, the scientific approach is based on laboratory experimentation using the parameters in Table 3.3.

It is observable from Table 3.3 that the parameters for assessing water quality from the scientific point of view are many and varied. On the other hand, traditional assessment is limited to only physical characteristics such as odour, colour, and taste.

Table 3.3 Scientific indicators of water quality

General	Temperature
	Oxygen (O_2)
	Oxygen saturation
	Suspended matter radioactivity
	Tritium
Inorganic compounds	Calcium (Ca^2)
	Magnesium (Mg^2)
	Total hardness
	Total dissolved salts
Organic compounds	Chemical oxygen demand (COD) mg/l
	Potassium permanganate mg/l
	Colour
	Odour number
Eutrophic compounds	Ammonia (NH_4)
	Nitrate (NO_2)
	Nitrate (NO_3)
	Total phosphate (PO_3 PO_4)
Metallic elements	Iron
	Manganese
Bacteria (microbiological)	Faecal coliforms
	Coliform organisms
	*Streptococcus faecalis**

Source: *Standard Methods.* AWWA/APHA/WEF (1992)
* *The Oxoid Manual* 7

Crop yield assessment

Yield has been defined as the amount of product available and useful for collection or harvest at a given point in time, i.e. that which can be used commercially (FAO, 2001). It can also be interpreted as the total biological potential of a species. Traditional assessment of yield is dependent on the type of crop. In the case of fruits, the assessment is based on fruit colour, and cutting and tasting of the fruit. In the case of other crops such as plantain, cassava, cocoyam, and maize, it is dependent on the type of crop variety and maturation period.

Maturation period

Table 3.4 summarizes how communities assess maturity of crops locally. This knowledge, combined with physical characteristics such as size and colour, is used to determine yield per crop. Total crop yield estimation is speculative. By contrast, the scientific approach uses a variety of methods for yield assessment. Some of them are indicated in Table 3.5.

Species diversity

Farmers' assessment of species diversity is based on three parameters, namely leaf type, colour of the tree stem, and branching of the tree.

Yield/maturity assessment

Farmers contend that no two trees have the same type of leaves, colour of the tree stem, and branching of the tree. On the other hand evenness of species is determined by counting species in various farms or, alternatively, by the amount of fuelwood collected by the farmer. The scientific method of measuring richness and diversity, e.g. that by Shannon-Weaver (1949) and Simpson (1949), is more systematic.

Table 3.4 Indicators used traditionally by communities to assess maturity of crops

Food crop	Indicator
Plantain	Tip of the finger becomes black, and/or browning of leaves around the pseudostem
Cassava	Cracking of the soil beneath and around the base of the stem/the bulging of the tubers
Maize	Development of black layer of the maize caryopsis or browning of the silk
Cocoyam	Browning of the leaves

Table 3.5 Examples of techniques used for quantifying product yield

Variable	Methodology
Fruit yield per season	Ground-level traps. Four isolated selected 15 × 1m^2 plots randomly located beneath crown. Number of intact, predated, immature, and mature fruit recorded every 7–10 days in plot
Fruit yield per season	Fruit counted *in situ* on sample trees at frequent (weekly) intervals. Counted fruit marked with paint to avoid repeated counts
Fruit, leaves, etc.	Randomized branch sampling. Branching pattern defined as numbered segments between branch nodes. Path from trunk to branch tip selected using random selection at each node. Fruit, leaf, etc. counts undertaken at distal end of path. Pooled results from several randomly selected branches are a non-destructive, precise, and statistically reliable method of estimating fruit yield of tree. There are several refinements of method, e.g. path selection proportional to size of available segments at a node, etc.
Leaves	Pipe model. Non-destructive regression technique for estimating leaf biomass and area from branch cross-sectional area. Pipe model based on observation that transpiration rate of canopy is proportional to leaf area, sapwood cross-sectional area, and conductivity of water-transporting tissue. Therefore size of stem is proportional to leaf mass and area. So leaf and area can be estimated by measurement of stem cross-sectional area (NB: needs to be very accurate – mm). Sample branches selected systematically to represent different branch heights. Regression analysis without constant

Table 3.5 (cont.)

Variable	Methodology
Palm leaves	All leaves measured. Partially open leaves counted as fraction of open leaf. Leaf length measured monthly to track growth
Palm stem increment	Leaf scars counted at monthly intervals. Stem growth quantified as height increment (cm) per leaf scar
Palm age	Count of leaf scars, assume constant rate of production to give estimates of age and numbers of years to reach critical heights
Bulb size	Measurement of maximum width of largest leaf on each plant. Regression analysis performed on a random sample of 50 plants at each site indicated that the largest leaf's maximum is strongly correlated to total leaf area. Total leaf area already shown to be an indicator of bulb size
Bamboo biomass	Measure clump dimensions on orthogonal axes at ground level, 1 m and full canopy extent. Map these as concentric ellipses. Determine biomass as volume of cone projected upwards from the base of the clump. Site index = clump volume/clump
Bushmeat weight	Opportunistic records of weights of captured animals in three villages used to supplement animal census

Source: FAO (2001)

Characteristics of traditional and scientific assessment techniques

There are observed differences in the assessment techniques. Traditional assessment techniques do not require extensive data gathering. Since observation is the key methodology, assessment is quick and monetary costs are very low. Decisions are made *in situ*, and the only guide is experiential knowledge. The other method of assessment is by comparison with previous crop performance. On the other hand, scientific assessment requires data gathering and measurements over a period of time and has a lot of cost implications.

However, both the traditional and scientific methods use physical characteristics, even though the parameters differ with respect to colour.

Soil management

The traditional approach to soil fertility management is seen in a number of ways, such as the application of the *proka* system of mulching (which is discussed in Chapters 6, 11, 14, and 19), the growing of cover crops, and fallow practices. Although burning is generally not an adopted method for soil conservation, the farmers are aware that most burning sites are fertile because the residual ash contains potassium, which is very good for maintaining soil fertility.

The scientific approach to soil management is the application of inorganic fertilizers, organic manure such as compost, mulching, growing of cover crops, and fallow.

Conclusion

The basic differences between traditional and scientific assessment are that whereas the traditional methodology is essentially qualitative and subjective, the scientific approach is quantitative and objective. Also the results of the scientific method tend to be more reliable and verifiable.

Whilst scientific data are often recorded and can be retrieved anywhere by the use of information technology, the traditional method suffers from time lapses and transmission losses. Consequently, over a long period, the traditional knowledge may be lost entirely.

In terms of monetary implications, the community approach is less costly. Both methods are susceptible to errors. As to the reliability of traditional assessment of natural resources, the margin of error is dependent on the experience of the farmer. In the case of the scientific approach, the errors are dependent on the quality of the information collected and the experiments conducted.

REFERENCES

American Public Health Association, *Standard Methods for the Examination of Water and Waste Water*, 18th edn, Washington, DC: APHA, 1992.

Bridson, Y., ed., *The Oxoid Manual*, 7th edn, Alton: Alphaprint, 1995.

Buabeng, S. N., "Socio-economic conditions and its effects on agricultural and biological changes in Tano Dumase and Jachie demonstration sites in central Ghana", unpublished manuscript, Bureau of Integrated Rural Development, Kwame Nkrumah University of Science and Technology, 1998.

Food and Agricultural Organization, *Non-Local Food Products Resource Assessment of Non-Wood Forest Products: Experience and Biometric Principle*, Rome: FAO, 2001.

Hountondji, P., ed., *Endogeneous Knowledge Research Trails*, Chippenham: Antony Rowe, 1997.

Shannon, C. E. and W. Weaver, *The Mathematical Theory of Communication*, Urbana: University of Illinois Press, 1949.

Simpson, E. H., "Measurement of diversity", *Nature*, No. 163, 1949, p. 688.

4

Farmer strategies of managing agrodiversity in a variable climate in PLEC demonstration sites in southern Ghana

Edward Ofori-Sarpong and Felix Asante

Introduction

In most regions of Africa, rainfall in the 1950s and 1960s was fairly plentiful and regular, but variability doubled in the 1970s and 1980s in some regions. Climate change results from interaction of a complex set of activities, including anthropogenic ones. These activities, in turn, are influenced by climatic change. The interrelationships are major subjects of research.

In Ghana as elsewhere, there is an increasing awareness and appreciation of environmental problems. However, because the relationship between climate change and its effects on agriculture is frequently non-linear, small changes in climate can have large effects.

This chapter discusses farmers' strategies of managing agrodiversity in the wake of climate change or variability with special reference to PLEC demonstration sites in southern Ghana (Map A).

Major characteristics of the ecological zone

There are three PLEC demonstration sites in southern Ghana, all of which are situated in the Eastern region, and within the forest-savanna zone (Maps B, C, D, E). Like the rest of Ghana, the Eastern region is situated on a dissected ancient plateau of metamorphosed and sedimentary rocks. Gyamfiase-Adenya is located in the Akuapem North district, Sekesua-Osonson in Upper Manya Krobo district, and Amanase-Whanabenya in Suhum-Kraboa-Coaltar district. Migrant cocoa farmers founded them in the twentieth century (Gyasi *et al.*, 1995).

Average annual rainfall ranges between 1,200 and 1,450 mm. It is bimodal and adequate for two crops in a year. The major cropping season is from April to July and the minor season from September to October, with the months of November to March being associated with the dry *harmattan* season.

Soils are predominantly ochrosols. The vegetation has modified from a much thicker true forest to secondary regrowth with a few pockets of thick forest (Thompson, 1910; Chipp, 1927; St Clair Thompson, 1936; Taylor, 1952, 1960; Hall and Swaine, 1976, 1981).

Despite growing monocultures there still is high agrodiversity. This reflects both the transitional nature of the ecosystem, which permits cultivation of crops adapted to humid and dry conditions, and the great ethnic and cultural diversity arising from migration.

Located in Akuapem district, cradle of Ghana's cocoa industry, Gyamfiase-Adenya is settled predominantly by native Akuapem people and Ewe migrant farmers on the basis of a mosaic landholding pattern.

In Sekesua-Osonson, the inhabitants are mainly the offspring of Krobo migrant cocoa farmers. They settled there on the basis of a patrilineal linear *huza* landholding arrangement.

Amanase-Whanabenya is settled by a mix of the offspring of migrant Akuapem and Siade/Shai people and a growing component of recent migrant Ewe settled farmers. Landholding arrangements are both the mosaic and linear *huza* type. Suhum-Kraboa-Coaltar district forms part of the historic southern Akyem cocoa frontier. As in Akuapem, the inheritance system is both matrilineal and patrilineal.

Farming by smallholders is the main occupation in all the sites. They grow mainly food crops, notably cassava, maize, plantain, cocoyam, and oil palm. Other economic activities include cassava processing, distilling of *akpeteshie*, a local gin, and extraction of oil from the fruit and kernel of the oil palm.

Evidence of climate variability

There is an evidence of considerable climatic variability in the study areas. The meteorological statistics are buttressed by claims of farmers with whom discussions were held. Very nearly all of them reported substantial changes in the climate, especially in volume and reliability of the rainfall, and intense sunshine.

Using four proxy rainfall stations, namely Akropong, Aburi, Koforidua, and Suhum (Figs 4.1–4.4) over a 30-year period, the trend displays fluctuations in the annual rainfall totals. The coefficient of variability over the period was 18.9 per cent for Koforidua, 20.3 per cent for Akropong, and 23.3 per cent for Aburi. These figures point at a moderate to high rainfall variability.

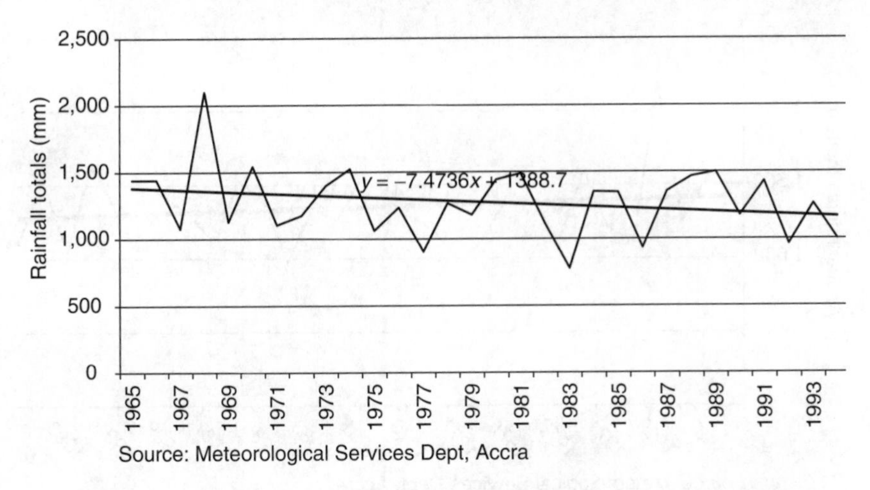

Figure 4.1 Annual rainfall fluctuations and trend at Akropong (1965–1994)

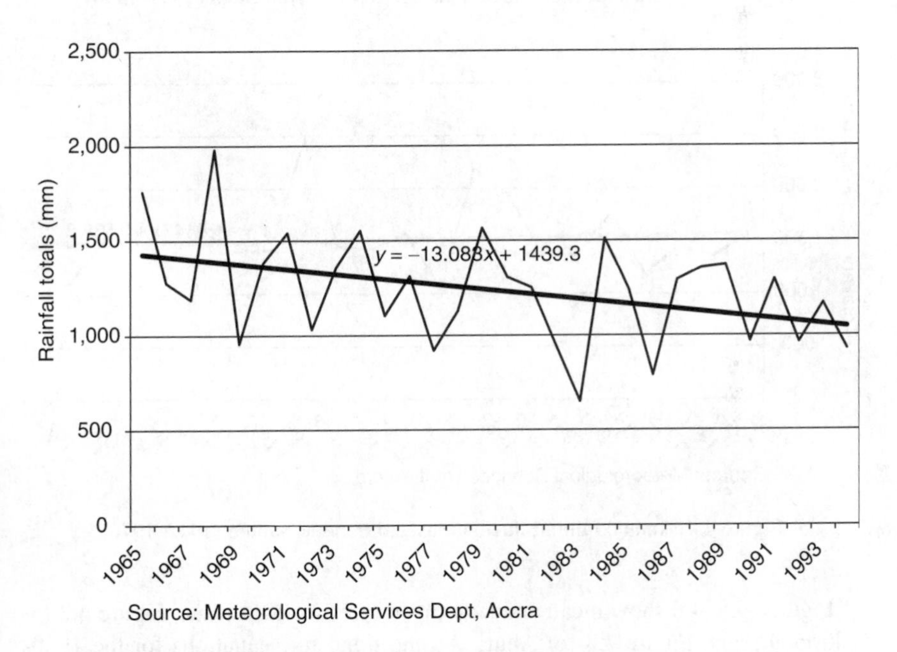

Figure 4.2 Annual rainfall fluctuations and trend at Aburi (1965–1994)

Apart from Suhum, all the other stations showed a negative trend line equation, which is indicative of a general decline in rainfall. Since 1980, most of the rainfall stations in Ghana have exhibited this declining pattern. If data were available for Suhum, they probably would have exhibited similar characteristics.

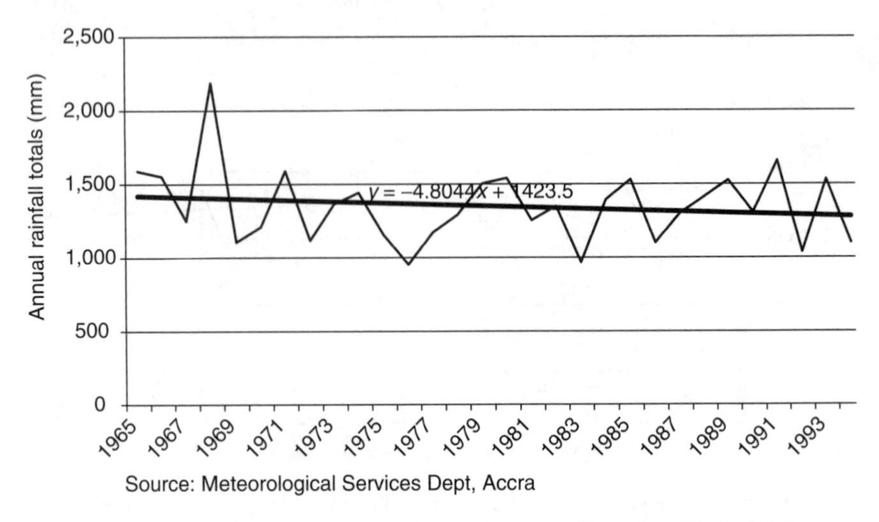

Figure 4.3 Annual rainfall fluctuations and trend at Koforidua (1965–1994)

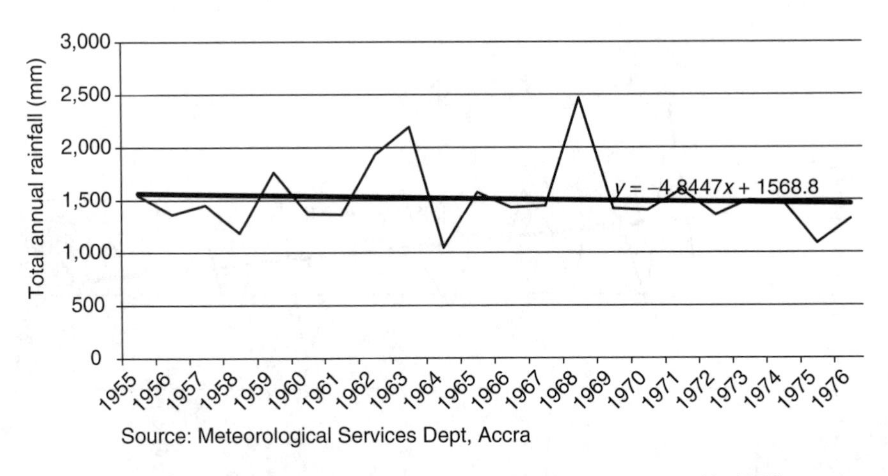

Figure 4.4 Annual rainfall fluctuations and trend at Suhum (1942–1976)

Figures 4.5–4.6 show mean annual temperatures for Koforidua, Figure 4.7 for Akropong, and Figure 4.8 for Aburi. All the trend line equations for the 1980s show positive x-values indicating rising temperatures. The national mean temperature has also changed from about 25 °C in the mid-1970s to about 27 °C in recent times (Benneh and Dickson, 1977). A comparison of rainfall and temperature for the three stations reveals an upward temperature trend and a downward rainfall trend. This may lead to increased evapotranspiration rates.

Figures 4.9 and 4.10 show the mean water balance for Koforidua over the periods 1972–1981 and 1991–1998. It was computed on the basis of mean monthly rainfall

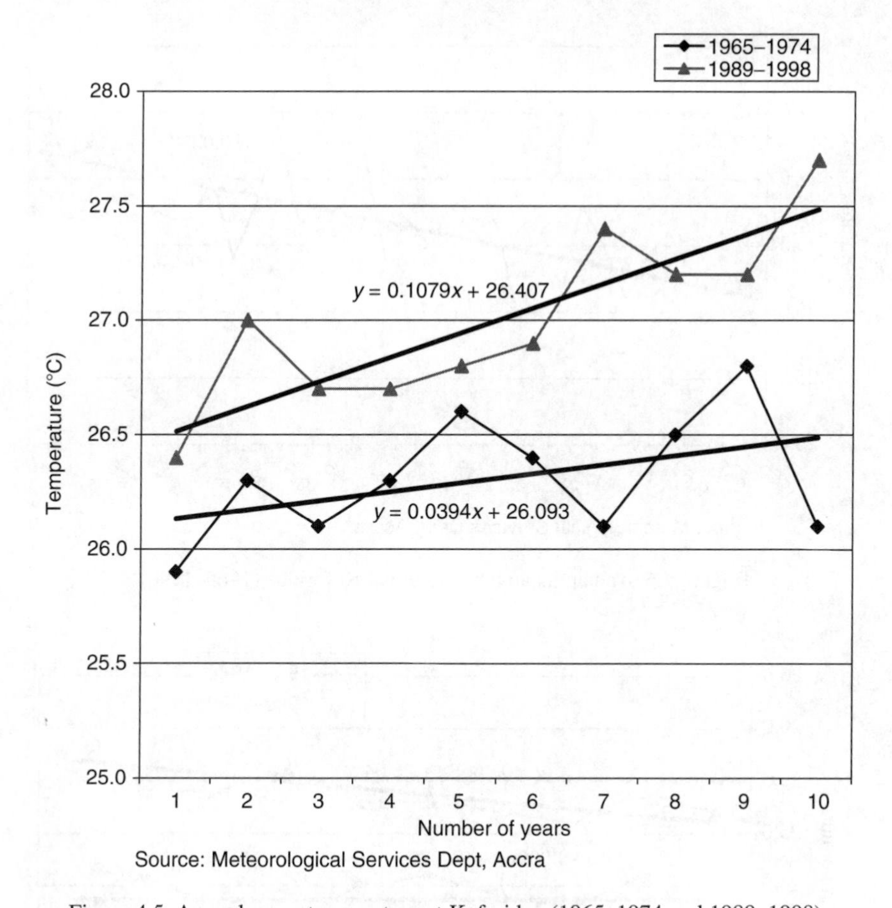

Figure 4.5 Annual mean temperature at Koforidua (1965–1974 and 1989–1998)

and potential evapotranspiration (PET). Although not representative of the entire southern Ghana demonstration site, the water balance makes it possible for one to establish fairly accurately the degree to which water requirements for general plant growth are satisfied in a given place. Crop possibilities can only be determined when the moisture requirements of individual crops are known (Benneh, 1971).

Comparing the mean water balance at Koforidua over the two periods, it is observed that there is now a distinct break between the major and minor planting seasons, and a higher PET for the 1990s. This is indicative of increasing water deficit and decreasing soil moisture availability for plant growth in the minor season farming because most crops have "moisture-sensitive periods" during which a water deficit could reduce the economic yields much more than at other periods.

The farmers who were interviewed generally saw a lot of evidence of climatic change. They observed a change in the natural vegetation and in the relative importance of some trees, for example the gradual disappearance of timber

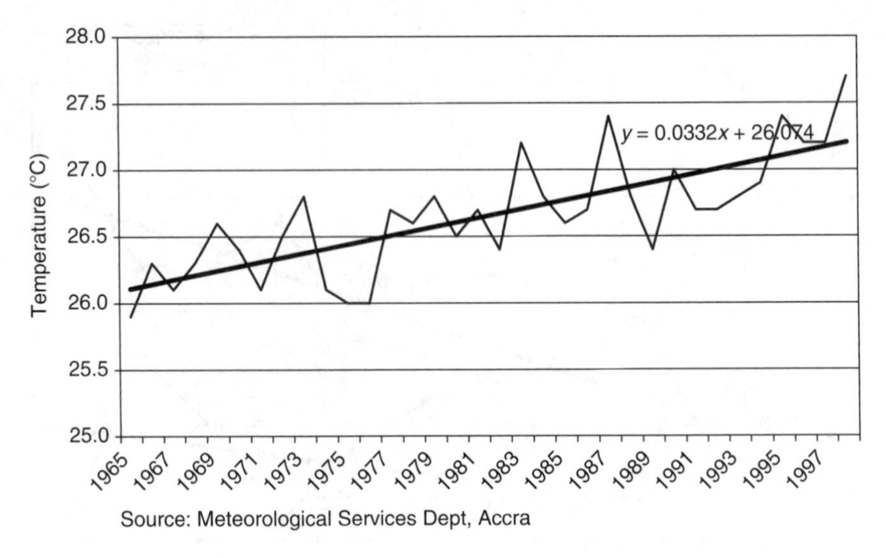

Figure 4.6 Annual mean temperature at Koforidua (1965–1998)

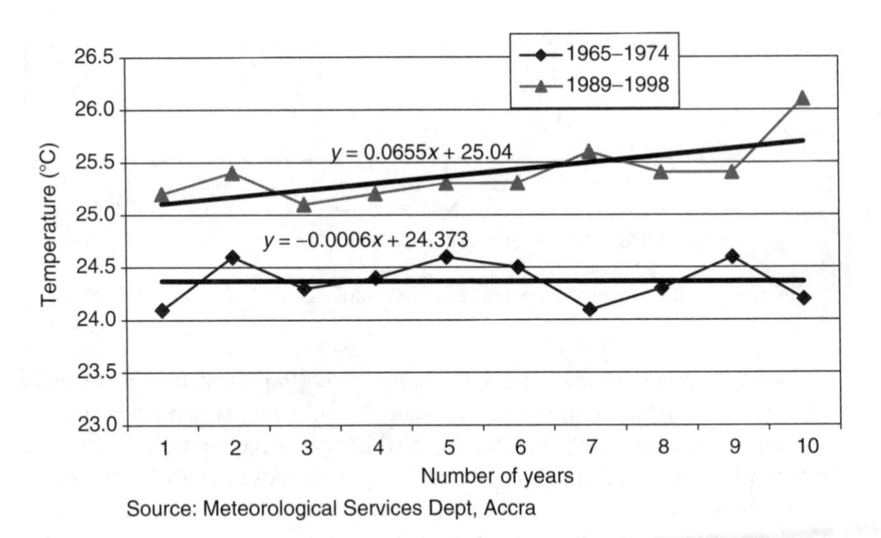

Figure 4.7 Annual mean temperature at Akropong (1965–1974 and 1989–1998)

species such as *emere*, *wawa*, and *odupko*. Others include lower rainfall reliability and a shift in the planting season. They also reported that the traditional signs of the start of rainy season are no longer reliable. Rivers dry up earlier in the dry season than in times past, and some hitherto seasonal rivers have dried up completely.

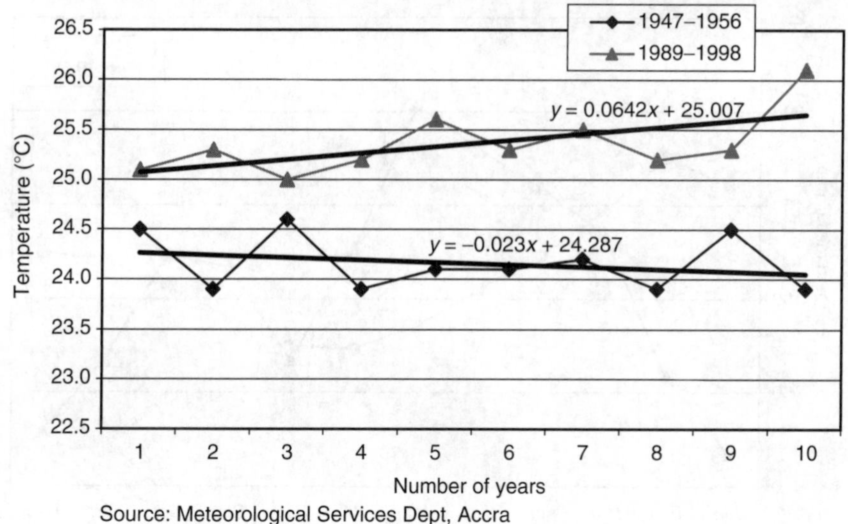

Figure 4.8 Annual mean temperature at Aburi (1947–1956 and 1989–1998)

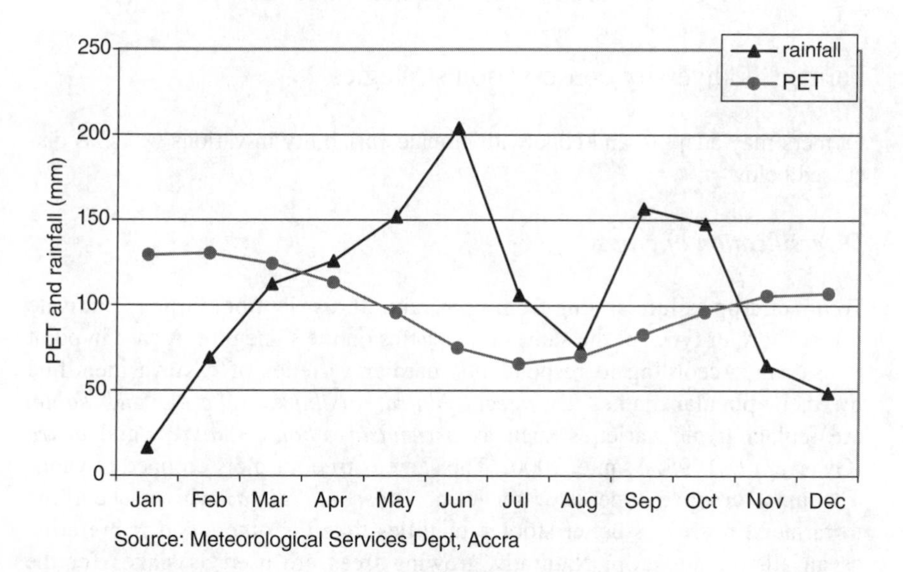

Figure 4.9 Mean water balance at Koforidua (1972–1981)

Although climatological evidence and local opinion suggest that rainfall has been increasingly variable and skewed towards lower annual totals since the 1980s, with an accompanying increase in temperatures over the same period, insufficient evidence exists to support the view that long-term climatic change is taking place.

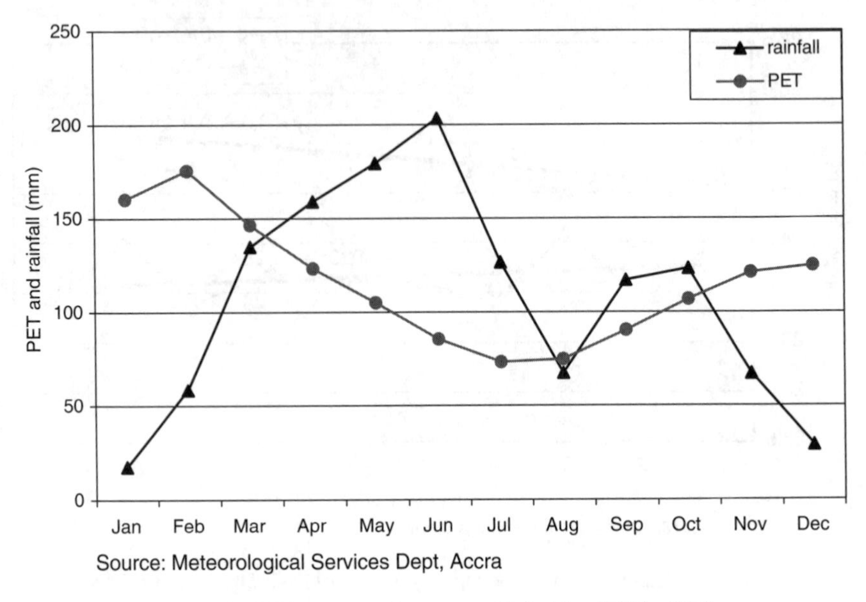

Figure 4.10 Mean water balance at Koforidua (1991–1998)

Farmer biodiversity conservation strategies

Farmers may adapt to and cope with climate variability in various ways, as discussed below.

Diversification of crops

To offset crop failure arising from poor rainfall distribution, farmers cultivate several hardier types of the same crop species on the same plot. A case in point is cassava. According to respondents, hardier varieties of cassava identified by their popular names as *agege*, *biafra*, *atsilapka*, *gbezey*, and *kable*, are replacing old varieties such as *asramnsia*, *abontem*, *tuaka*, and *ankra* (Gyasi *et al*., 1995; Asante, 2000). The same is true for plots cropped to yams. The most cropped type is water yam, *Dioscorea alata*, which, according to farmers, possesses better storage qualities than the others and is preferred as an all-weather crop. Naturally growing trees are used as stakes for the yams. In the Sekesua-Osonson demonstration site for instance, *nyabatso*, *Neuboldia laevis*, is the dominant tree used for staking the yams. Apart from being medicinal, *nyabatso* is reputed to enhance soil fertility and result in large yam tubers. Through exchange of visits between and within demonstration sites, the practice of using *nyabatso* for yam cultivation has stimulated on-farm *in situ* conservation of trees by farmers in the demonstration sites.

Nyamedua, Astonia boonei, *agyama, Alcornea cordifolia*, and *osese, Hollarrhena floribunda*, which, according to farmers, use their long tap-root system to draw water for plants in their environs, featured prominently in Amanase and Gyamfiase as trees that combine well with crops (Asante, 2001).

Similarly, a diversity of vegetables like pepper, garden eggs, and okra is cultivated as a hedge against risks associated with drought. Attempts at finding the diversity index for the major staples, notably cassava and yams, were frustrated by difficulty in classifying them by their different local names, even within the same locality.

A typical cropping sequence begins by interplanting maize with minor crops including vegetables. This mixture is later interplanted with cassava and others such as banana or plantain. After two or three years of cultivation on the same plot, the plot is left to fallow. Plants normally left on farms as they fallow, and which are harvested from time to time, include perennials, notably tree crops, vegetables such as pepper and garden eggs, and annuals such as cassava and yams.

Crop biodiversity provides the ecosystem resilience necessary to cope with periodic stresses on the environment, such as drought and climate variability among others. These adaptive strategies bear testimony to the indigenous and sophisticated adaptation of production systems by farmers to the unique characteristics of their environment.

Planting materials

Farmers also respond to rainfall variability and unpredictability by using seeds saved from the previous harvest. These are obtained from other farmers at a token fee, and from the local markets. Agricultural extension shops are the purchasing points of high-yielding and early-maturing seeds of both cereals and vegetables.

The continuous use of seeds saved from previous harvests ensures maintenance of local genetic diversity available to farmers, thereby enabling them to manipulate production systems to suit local conditions better.

Tastes and preferences of the farmers to a large extent determine what they grow, and this also conditions their choice of planting materials collected from colleagues. The rationale for using seeds saved from the previous harvest is that such seeds are likely to survive any adverse weather during the current planting time because they survived the previous season. As noted earlier, several new cassava varieties have been introduced into the study area from neighbouring Togo and Benin. Notable among them are *gbezey*, *agbeli-atsilapka*, *yevuvie*, and *nyonuvie* (all Ewe names), which have been introduced in Gyamfiase and Amanase where there are a lot of tenant Ayigbe farmers. Through exchange of visits between demonstration sites and germplasm among farmers, these newly introduced varieties have diffused to Sekesua.

High-yielding improved maize types such as *obaatanpa* (Twi name) and *abelehie* (Ga-Adangbe name), though a bit late in maturing (between 90 and 110 days), are planted mainly in the major season, and early maturing types (between 64 and 75 days) like *abrowtia* (Twi name), ideal for the lean season, are planted in combination during the planting season. This strategy helps farmers to reap at least some maize for home consumption even in an unfavourable planting season. Even so, farmers still prefer the local type because it is tastier.

Local vegetable seeds for future planting are extracted from the ripe fruit, sun-dried, and stored in rags, paper, or bottles and normally kept in safe places to prevent insect attack. Okra and bean types are kept dry in their pods and kept until the time of their sowing. According to farmers the local types are adapted to their environment and are planted in combination with the improved types.

In addition, crossbreeding, especially of animals, is carried out. For instance, a well-built ram or billy goat could be borrowed from its owner during the mating season at a token fee of 5,000 cedis (approximately US$1) to mate with several ewes or nanny goats of the one who desires to have sheep or goats of that kind. In the process, animals well adapted to the dynamic environment are produced. This selective breeding practice enables the farmer to manipulate several indigenous species on the same plot, thereby encouraging a shift away from monoculture, associated with the use of improved varieties.

Crop conservation

A change in the types of crops grown is another biodiversity conservation strategy employed by farmers. They know that relying solely on a single crop is not safe. Therefore they diversify the types of crops they grow. More citrus, cashew, mango, and oil palm are being planted, as farmers now perceive that the soil is no longer good for cocoa, once a dominant crop but which is no longer grown extensively.

However, the choice of a particular crop is based on a farmer's knowledge, food preference, and consideration of yields and sale. These crops, according to farmers, are not drastically affected by weather variability once they mature and, as such, are used as a form of insurance.

Other crops include pineapple, green pepper, and cabbage.

These shifts can affect the production of staples. In the short term, agricultural production may be lower due to decreased land area for cultivation as some plots previously used for staple crops may be channelled into these diversified crops. All the same, the revenue accruing to households from the sale of these crops and others with ready markets can enable such households to buy food and increase their food security.

Home gardening

Home gardening, which is also common in the study area, is another means of conserving biodiversity. The home gardens contain a wider diversity of crops than some main farms. In some of the gardens, trees and crops which are reported as becoming rare or near extinction are found. Most of the crops grown are roots and tubers, notably yam, *Dioscorea*. As described by Gyasi in Chapter 5, and by Enu-Kwesi, Gyasi, and Vordzogbe in Chapter 12, in Sekesua-Osonson and other areas settled by offspring of migrant Krobo farmers, the combination of crops mimics the traditional agroforestry. The trees are used as stakes for the yams.

The general crop combination in the home gardens include peppers, other condiments, leafy vegetables that are in regular demand by the kitchen, plantain and bananas, and fruit trees such as mango, citrus, avocado, and sour sops. The combination of crops with trees maintains ecological stability through, among other things, minimization of soil erosion by canopies formed by the trees. It also ensures an effective use of the different soil nutrients by the different crops, and a balanced diet through the diversity of crops.

Other activities encouraging biodiversity conservation

There are more or less economic activities that generate more value from conserved biodiversity, or that generate income in some other ways that encourage conservation. They include snail rearing, plant nursery establishment, and beekeeping.

Snails, a prized source of meat that are found in the wild under humid conditions, have come under threat by overfarming and habitat destruction. In all three demonstration sites, some farmers are enriching biodiversity and enhancing income prospects by raising snails on a semi-intensive basis in their homes and in secondary forests conserved in their backyards. Although this begun on a small pilot basis with PLEC support in all the demonstration sites, it has been taken up by several individual PLEC farmers.

Plant nurseries are also operated in all the demonstration sites. They are owned on an individual private basis, and on a group basis by farmers' associations. Sale of the assorted seedlings from the nurseries yields income to supplement farm income. Additionally, PLEC farmers use some of the seedlings to rehabilitate deforested areas in their various localities. This helps to enrich the flora through integration of new plants, improve soils through increased biomass, and check erosion through more trees. Examples include reforestation around the Kaja waterfalls at Prekumase and Bukunya falls at Bormase, all in Sekesua-Osonson, and integration of a variety of seedlings from the nurseries into food crop farms, thereby enriching biodiversity.

At Homeso in Amanase-Whanabenya demonstration site, some of these seedlings form the basis of an arboretum being developed by the association of PLEC farmers in collaboration with the National Centre for Scientific Research into Plant Medicine.

There is also a growing popularity of the PLEC-supported practice of using home gardens and forests conserved near homesteads to keep bees for honey and wax. The bees derive nectar from the trees for their honey and this, in turn, fosters pollination of the trees. This is especially so in the Sekesua-Osonson demonstration site where the practice has evolved from the use of traditional earthen pots as hives to the use of modern wooden beehives. Prospects are promising for further expansion in the other demonstration sites where PLEC is encouraging popular awareness of beekeeping as a means of improving livelihood. The initial small-scale beekeeping has expanded to cover many more farmers within the demonstration site with both material and financial support from Heifer Project International, an international NGO.

Additional biodiversity conservation strategies include sacred groves of forests (as in Gyamfiase), taboos, and reverence for certain plant species. Like other land uses of their kind, sacred groves are protected and improved by vigilance, sanctions, firebelts, buffer zones, and replanting. In the past few years these have been propagated with PLEC support.

Conclusion

There is sufficient evidence of significant climatic variability in the study areas. The climatic analysis showed that there is a general decline in the amount of rainfall received, while temperature is on the increase. These two trends have affected the water balance of the areas, leading to moisture deficit. To overcome the problem of moisture deficit, early-maturing crops are cultivated.

Farmers have several indigenous ways of predicting the suitability or otherwise of an impending planting season. They also have various strategies for countering environmental degradation.

These strategies are manifested by the diversity of resource management and cropping systems, which in turn are based on indigenous knowledge of management of the fragile and variable environment, local genotypes of food crops, intercropping, and agroforestry systems. Other strategies include beekeeping, snail raising, and plant nursery establishment.

REFERENCES

Asante, F., "Adaptation of farmers to climate change: A case study of selected farming communities in the forest-savanna transitional zone of southern Ghana", unpublished, M.Phil thesis submitted to the Department of Geography and Resource Development, University of Ghana, Legon, 2001.

Benneh, G., "Water requirements and limitations imposed on agricultural development in Northern Ghana", in I. M. Ofori, ed., *Factors of Agricultural Growth in West Africa*, Legon: ISSER, 1971, pp. 71–80.

Benneh, G. and K. B. Dickson, *A New Geography of Ghana*, Metricated Edition. London: Longman Group, 1977.

Chipp, T. F., *The Gold Coast Forest: A Study in Synecology*, Oxford Forestry Mem. 7, 1927.

Gyasi, E. A., G. T. Agyepong, E. Ardayfio-Schandorf, L. Enu-Kwesi, J. S. Nabila, and E. Owusu-Bennoah, "Production pressure and environmental change in the forest-savanna zone of southern Ghana", *Global Environmental Change*, Vol. 5, No. 4, 1995, pp. 355–366.

Hall, J. B. and M. D. Swaine, "Classification and ecology of closed-canopy forest in Ghana", *Journal of Ecology*, Vol. 64, 1976, pp. 913–951.

Hall, J. B. and M. D. Swaine, *Distribution of Ecology of Vascular Plants in a Tropical Rainforest: Forest Vegetation in Ghana*, The Hague: Dr. W. Junk Publishers, 1981.

St Clair Thompson, G. W., *Forest Conditions in the Gold Coast*, Imperial Forestry Institute Paper 1, 1936.

Taylor, C. J., "The vegetation of Gold Coast", *Bulletin*, *Gold Coast Forestry Department*, 4, 1952.

Taylor, C. J., *Synecology and Silviculture in Ghana*, Edinburgh: Nelson, 1960.

Thompson, H. N., *Gold Coast: Report on Forests*, Colonial Report Miscellaneous, No. 66, 1910.

5

Demonstration sites and expert farmers in conservation of biodiversity

Edwin A. Gyasi

Introduction

The idea of demonstration sites where biodiversity management systems are developed and demonstrated, primarily on the basis of smallholder farmer knowledge, is central to the PLEC mission of finding optimal paths towards conservation and, ultimately, improvements in farmer livelihoods. Demonstration sites developed on the basis of expert farmer knowledge have a special significance because they provide a cost-effective complement to the essentially top–down conventional research, extension, and development approaches, especially in the rural areas of developing countries.

This chapter elucidates the notion of a farmer-based demonstration site and how it is being developed through PLEC work with special reference to Ghana.

The expert farmer concept

The many smallholder farmers who manage a substantial proportion of global biodiversity display varying degrees of understanding of and adaptation to the variable ecological, economic, and socio-cultural forces shaping that diversity. To this extent, all of them may be said to be farmer experts or biodiversity experts in their own right.

However, some farmers excel in their knowledge of biodiversity and, on top of this, use that knowledge effectively to solve production problems and maintain livelihoods on a sustainable basis. Relatively few in number, these outstanding agricultural actors are the real expert farmers who constitute the keystone, the core social catalysts in demonstration site activities (Padoch and Pinedo-Vasquez, 1999; Pinedo-Vasquez, Gyasi, and Coffey, 2002). They may be spotted or identified by "combining skilled field observations with the close participation of many [other] farmers; long-term research with frequent consultation with farmers'organizations" (Padoch, 2002: 104).

Demonstration site defined

Abdulai *et al.* see a demonstration site as:

> a place where PLEC scientists, farmers and other environmental stakeholders carry out work in a participatory manner to conserve and even enhance agricultural and biological diversity and the biophysical resources underpinning it. It is an area where the scientists work with farmers in the creation of projects that are [the farmers'] own and [where, together, the scientists and farmers] demonstrate the value of locally developed techniques and technologies. It belongs to the farmers, in that the work done in a demonstration site is the farmers' own. The role of scientists is only to facilitate, measure and evaluate local methods and help to select the method most likely to be sustained (Abdulai *et al.*, 1999: 19)

The definition goes on to say that "The sites may vary in size and a site may contain sub-sites, notably a farm or a patch. However, sites must be at the local or perceptible landscape level" (Abdulai *et al.*, 1999: 20)

In Ghana, the PLEC operational definition of the primary demonstration site is a smallholder farmer area measuring approximately 10×10 km = 100 sq. km. Such an area is small enough to facilitate focused in-depth fieldwork, but large enough to show significant internal agro-ecological variations and to permit study by aerial photographs and satellite imagery. Specific focal sites of demonstration activities lie within this area.

It is implicit in the definition that demonstration sites could assume different sizes, or occur at various spatial or geographical scales/levels of resolution.

Any of the following landscape levels, which are presented in a more or less hierarchical order, would seem to be appropriate for PLEC demonstrative activities.

- plot or patch within a farmed area
- farm, or a contiguous area managed as an agricultural unit
- agricultural holding, or a collection of farms managed as one entity
- farmstead, or household together with its associated landholding which, because it is the basic production and consumption unit, PLEC recommends as the primary focal point for the research work
- a cluster of farmsteads or of households constituting a farming village
- a set or cluster of villages covering a discernible sizeable area.

At each scale, the demonstration may be carried out in the context of the predominant land-use types or stages, such as any of the following ones proposed by Zarin, Huijin, and Enu-Kwesi (1999) for plant species diversity assessment:

- annually cropped units
- agroforests
- fallows (grass-dominated, shrub-dominated, and tree-dominated)
- orchards
- native forests
- house or home gardens
- edges
- any other appropriate types.

Key steps in the process of developing the demonstration sites

From the experience in Ghana, the first step is the identification of a potential demonstration site for the study on the basis of the following parameters:

- richness of existing agricultural and biological diversity
- extent of threat to the diversity
- level of documented knowledge about the site.

If the agrobiodiversity is found to be rich, but under threat, then further studies of the agro-ecological conditions may proceed, using as a starting point whatever documented information there might be.

In the initial investigative studies, as in subsequent applied work, it is important to involve local people, especially the chiefs, other leaders, and expert farmers who possess good knowledge of the agro-environmental history and situation in general. These key personalities may later serve as facilitators in any applied work.

An important initial step is participatory mapping of settlements with involvement of local farmers, which, among other things, may serve as a basis for the creation of a geographical information system (GIS) towards a standardized database for demonstration sites.

Investigative studies are followed by applied work that may involve any of the following:

- consolidation of local contacts, including formalization of selected local people as facilitators, or expert farmers for demonstrations
- dialogue through group discussion with farmers to determine relevant needs on a prioritized basis
- organization of farmers and other local users or managers of the land resources into farmers' associations to facilitate popular PLEC scientist-farmer interactions
- promotion of farmer-led conservation measures
- gradual collaboration with agricultural extension agents and other appropriate governmental and non-governmental agents.

By extrapolating from the PLEC global experience, Pinedo-Vasquez, Gyasi, and Coffey (2002) have developed the prototype of a more or less universal model of the PLEC demonstration approach towards conservation and development. As captured by Figure 5.1, the approach proceeds according to the following sequence:

- formation of a team composed of a mix of relevant specialists, most notably scientists, technicians, and farmers
- farmer-based site assessment that is informed by existing conservation and development initiatives
- inventorying and documentation of conditions relevant to agriculture and biodiversity

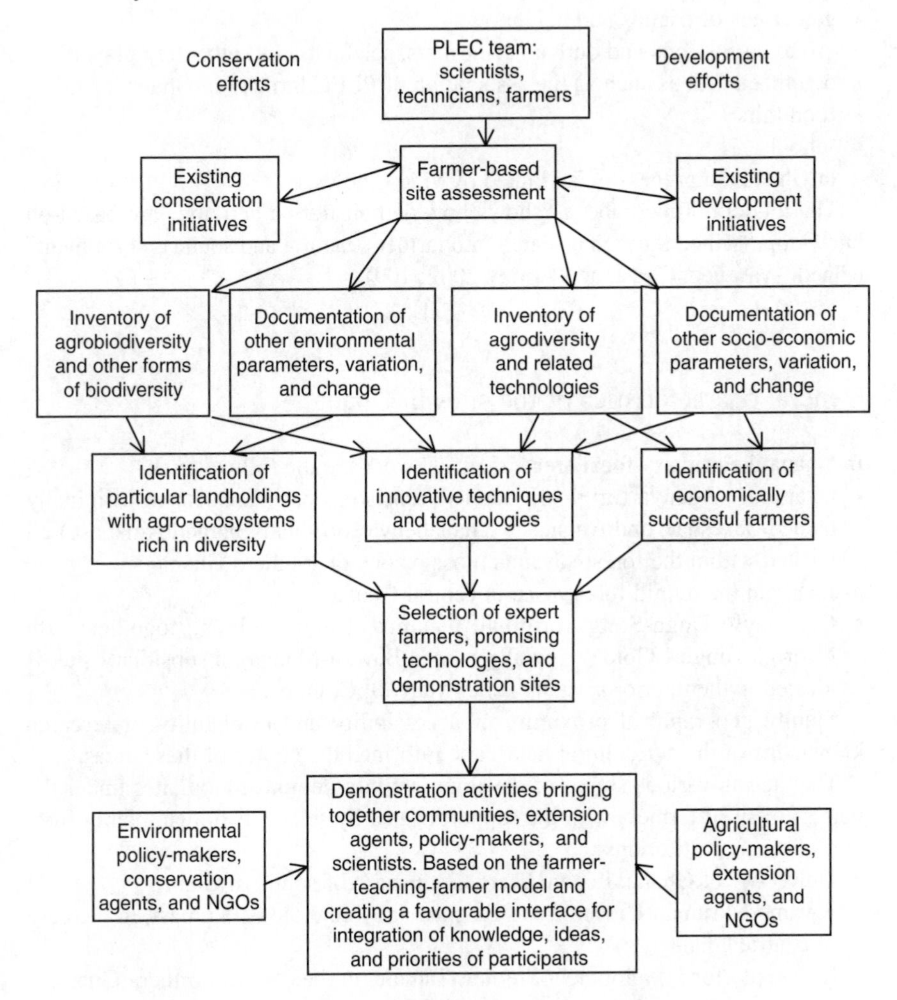

Source: Pinedo-Vasquez, Gyasi, and Coffey (2002)
Figure 5.1 The PLEC demonstration approach

- identification and selection of instructive management techniques and their practitioners, the expert farmers
- actual demonstrations and dissemination led by farmers, with scientists playing a monitoring and supportive role in collaboration with policy agents and other stakeholders.

Among the diversity of ways of identifying appropriate management systems or agrotechnologies and of demonstrating and disseminating them are the following:

- farm visits
- family reunions
- gatherings of friends and neighbours
- group discussions and durbars of farmers, scientists, and other key players
- organized groups such as the association of PLEC farmers in Ghana
- food fairs
- schools.

In Ghana the process is facilitated by video shows.

The model outlined above "shows how demonstration activities are based on knowledge gained through research into farmers' natural and social environment" (Pinedo-Vasquez, Gyasi, and Coffey 2002: 109).

General characteristics of the sites in Ghana

In Ghana the primary focal areas of PLEC work are the following:

- Gyamfiase-Adenya (originally named Yensiso), Sekesua-Osonson (originally named Sekesua), and Amanase-Whanabenya (originally named Amanase) all located within the forest-savanna mosaic zone of southern Ghana
- Jachie in the humid forest zone in central Ghana
- Bongnayili-Dugu-Song (originally named Dugu), which, together with Nyorigu-Binguri-Gore (originally named Bawku-Manga), a subsidiary site, is located in the interior savanna zone (Maps B, C, D, E).

Mainly geographical proximity or accessibility and availability of research knowledge of the agricultural landscape informed the choice of these areas.

They are in various stages of development into demonstration sites under the management of farmers with the support of nearby scientists drawn mainly from the following institutions:

- University of Ghana, Legon/Accra, in the case of southern Ghana
- Kwame Nkrumah University of Science and Technology, Kumasi, in the case of central Ghana
- University for Development Studies, Tamale, in the case of northern Ghana.

A major characteristic of the sites is the considerable pressure exerted on biophysical resources by the predominantly agricultural population through cropping, grazing, wood harvesting, settlements, and the sheer numbers of people (Gyasi *et al.*, 1995; Gyasi and Uitto, 1997).

Activities

In Ghana, work in the demonstration sites progressed systematically from the initial focus on understanding baseline agro-ecological conditions and on establishing social contacts, through the following stages:

- assessment of biodiversity and agrodiversity
- identification of traditional, indigenous, or local systems of managing and conserving biodiversity, especially in an agricultural context
- conservation promotion
- promotion of economic activities that motivate farmers to conserve.

All these are towards the general PLEC quest for models of conserving biodiversity.

Outputs include essentially traditional farm management practices that conserve biodiversity (Table 5.1). Among them is traditional agroforestry, which involves cropping among trees left *in situ* in farms (Plate 1). This has the advantages of:

- conserving trees while producing a diversity of food and other crops
- regenerating soil fertility through nitrogen fixed by some of the plants and by the substantial biomass they generate.

Another example, as discussed in some detail in Chapters 11 and 14, is the no-burn farming that avoids use of fire for clearing vegetation, and which involves mulching by leaving slashed vegetation to decompose, in the practice called *oprowka* or *proka* by Akan-speaking people (Plate 2). It maintains soil fertility by conserving and stimulating microbial activity and by humus addition from decomposing vegetation. Propagation of plants, including farm seeds in the soil, is promoted by this method.

Through associations of PLEC farmers, which involve expert farmers, these and other management practices form a basis of PLEC experiments and demonstrations in biodiversity conservation (Table 5.2). Foremost among the experiments is one designed to determine the scientific basis of the popular claim by farmers that certain trees combine effectively with food crops, while others do not (Owusu-Bennoah and Enu-Kwesi, 2000; see also Chapters 13 and 14). Another is the demonstration of how income may be generated from, or value added to, conserved resources through the use of home gardens, fallow areas, and forest maintained in the backyard for keeping bees for honey and wax (Plate 3; see Chapters 12 and 17). Beekeeping has caught on the most in Sekesua-Osonson, where it involves approximately 70 households (Gyasi and Nartey, 2003). A further example of a value-added activity is

Table 5.1 Essentially traditional management practices/regimes in PLEC demonstration sites in Ghana

Practices/regimes	Major advantages
Minimal tillage and controlled use of fire for vegetation clearance	Minimal disturbance of soil and biota
Mixed cropping, crop rotations, and mixed farming	Maximize soil nutrient usage, maintain crop biodiversity, spread risk of complete crop loss, enhance a diversity of food types and nutrition, favour soil regeneration
Traditional agroforestry, cultivating crops among trees left *in situ*	Conserve trees; regenerates soil fertility through biomass litter. Some trees add to productive capacity of soil by nitrogen fixation
Proka, a no-burn farming practice that involves mulching by leaving slashed vegetation to decompose *in situ*	Maintain soil fertility by conserving and stimulating microbes, and by humus addition of decomposing vegetation; conserves plant propagates including those in the soil
Bush fallow/land rotation	A means of regenerating soil fertility and conserving plants in the wild
Usage of household refuse and manure in home gardens and compound farms	Sustains soil productivity
Use of *nyabatso*, *Newbouldia laevis,* as live-stake for yams	The basically vertical rooting system of *nyabatso* favours expansion of yam tubers, while the canopy provides shade and the leaf litter mulch and humus. It is thought that *nyabatso* fixes nitrogen
Staggered harvesting of crops	Ensures food availability over the long haul
Storage of crops, notably yams, *in situ* in the soil for future harvesting	Enhances food security and secures seed stock
Conservation of forest in the backyard	Conserves forest species, source of medicinal plants at short notice, favours apiculture, snail farming, and shade-loving crops such as yams

Source: PLEC fieldwork since 1994

the processing of cassava into flour for bread and pastries by PLEC female farmers (Gyasi, 2001).

The associations of PLEC farmers that facilitate these activities do so by serving as a medium for:

- farmer-scientist interactions and collaborative work
- farmer-to-farmer interactions, including exchange of knowledge and germplasm

Table 5.2 Demonstration activities in Ghana

Activities	Outcome	Location
Experiment to determine trees that combine or do not combine well with food crops, and to determine optimal tree-food crop spacing	• Still under assessment, but initial findings appear to be generally consistent with claims by farmers	Duasin and other locations in Gyamfiase-Adenya
Use of selected farms as agroforestry models	• Increased popularity of traditional agroforestry • Apparent improvement in soil fertility and crop yield • Increased fuelwood	All sites in southern and central Ghana
Use of home gardens as germplasm bank and source of food, medicinal, and other useful plants	• Spread of home gardening • Reported income increase • Growing modelling of school gardens on home garden principles	All sites in southern Ghana
Yam management: techniques of planting, staking, harvesting; storing *in situ* in the soil unharvested small yams for use as seeds	• Spread of yam farming involving a diversity of varieties	Initiated in Gyamfiase-Adenya, but spread to other sites in southern Ghana
Conservation of over 20 varieties of yam in a demonstration plot at Dugu	• Conservation of a diversity of yams including disappearing ones • Propagation of rare yams among farmers	Bongnayili-Dugu-Song northern Ghana
Conservation of sacred forest groves through PLEC farmer associations	• Conserved assorted trees and diversity of other plants • Popular awareness of conserving biodiversity through conserved forest	All sites in Ghana
Medicinal plant conservation through arboreta	• Conserved assorted medicinal plants, which are starting to yield a modest income • Popular awareness of methods and prospects conserving medicinal plants through arboreta	Amanase-Whanabenya and Sekesua-Osonson, southern Ghana

Table 5.2 (cont.)

Activities	Outcome	Location
Model systematic biodiverse farming	• Inspired integration of trees into farming around the Adenya to Gyamfiase road along which the model farm is located	Gyamfiase-Adenya, southern Ghana
Demonstration of plant propagation by grafting/budding and spilt-corm techniques	• Over 100 farmers had learnt the technique from a PLEC-sponsored training programme at the University of Ghana Agricultural Research Station (ARS) or, subsequently, through farmer-to-farmer demonstration An estimated 40 of them were practising	All sites in southern Ghana
Pilot plant nurseries through PLEC farmer associations	• Establishment of similar group-owned nurseries in all sites, with seedlings output sold to sustain work of the farmer associations • Inspired privately owned nurseries operated on commercial basis • Propagation of rare exotic and endemic plant species	Initially limited to southern Ghana, but now spread to all sites in the country
Biodiversity conservation through multi-purpose floral and faunal nursery	• Female farmers trained in aspects of biodiversity management, notably snail farming, beekeeping, and plant nurseries, for income	Jachie, central Ghana
Experiment in semi-intensive commercial breeding of the giant African snail, making use of the canopy of a giant tree in a conserved forest patch at Obom	• Growing snail population • Demonstration of commercial value awaiting significant increase in snail population	Gyamfiase-Adenya, southern Ghana
Semi-intensive commercial raising of rare local breeds of domestic fowl	• Rapid multiplication of fowl • Popular awareness that certain breeds are getting rare, hence a need for their conservation • Growing awareness of the semi-intensive method as opposed to the common	Jachie, central Ghana

	• Commercial value yet to be systematically assessed	
Biodiverse cropfarming using traditional and modern principles	• Increase in farmers' income from the diversity of crops raised • Principles integrated into farming by others, including schoolchildren managing school gardens	Gyamfiase-Adenya, southern Ghana
Conservation of rare indigenous varieties of rice by female expert farmers, for domestic consumption and commercial purposes	• Conservation and propagation of disappearing varieties of crop • Improved farmer income from sales	Nyorugu-Benguri-Gore
Forest conserved in the backyard for beekeeping (for honey and wax) as a means of generating income	• Remarkable spread of beekeeping involving over 70 farmers. This development has attracted substantial financial support from a Ghanaian affiliate of an American NGO	Initially at Sekesua-Osonson, but has now spread to other sites in southern and northern Ghana
Development of teak woodlot on a commercial basis	• Popular awareness of prospects of income from woodlots • PLEC female farmers of Sekesua-Osonson plan similar trial • Actual commercial viability yet to be assessed	Jachie, central Ghana
Integration of high-yielding citrus into traditional farming	• Initial group-owned citrus have started flowering and bearing fruits	All sites in southern Ghana
Spinning and weaving of cotton by elderly PLEC females for benefit of younger women	• Acquisition of spinning and weaving skills by young females and potential reduction of unemployment, poverty, and out-migration	Bongnayili-Dugu-Song, northern Ghana
Processing of cassava into flour for bread and pastries by PLEC female farmers	• Improved rural incomes	Jachie, central Ghana
Piggery and sheep raising to be integrated into the conservation process on a commercial basis	• Activity still in formative stage	Amanase-Whanabenya and Gyamfiase-Adenya, southern Ghana

- reaching out to farmers and sensitizing them to issues of conservation and development
- mobilizing the latent knowledge, energy, and other resources of farmers for the purpose of conservation and development
- tapping or accessing external support for farmers
- carrying out demonstrations
- in general, empowering farmers politically, socially, and economically.

The capacity of the associations is strengthened by bank accounts opened by them on the advice of PLEC, and by links developed with government as well as non-governmental organizations.

Expert farmers play a central role in the associations. They are identified or spotted by:

- seeking views of farmers as to whom they consider to be exceptionally knowledgeable in various areas of resource management, notably conservation of particular species of crops, soils management, and identification, utilization, and conservation of medicinal plants.
- observing and monitoring how a farmer actually manages biophysical resources in the field
- listening to a farmer's stories and impressions about natural conditions, and how they relate to agriculture.

Expert farmers are exemplified by the following:

- Emannuel Nartey, an expert in apiculture (Plate 4)
- Odorkor Agbo, who excels in management of assorted yams within an agroforestry system (Plate 5)
- Cecilia Osei, who excels in the *oprowka/proka* no-burn system of farmland preparation (Plate 6)
- George Amponsah Kissiedu, whose biodiverse home garden, developed on the basis of both traditional and modern farming principles, inspires similar gardens including one established by schoolchildren (Plate 7).

Conclusion

It appears feasible to use the demonstration and expert farmer concepts as a strategy for developing optimal methods for conservation, especially of biodiversity within agriculture, and through it improvements in farmer livelihoods. A measure of feasibility is the increasing success of PLEC initiatives and of growing self-reliance of the pivotal PLEC farmers' associations that draw inspiration from the knowledge of expert farmers in PLEC demonstration sites in Ghana. Commitment of schoolchildren is another measure of feasibility and of the promise of sustainability of the PLEC initiatives. It is echoed by the following poem titled "The Trees Prayer" by a young pupil at Tinkong, a village near the Gyamfiase-Adenya site:

Every year many trees are cut all over the country for timber, firewood, charcoal, farming activities and more. Have you ever stopped to think about the importance of trees to man?

Just look around you. You will find several uses of trees (wood). Think about many things in your schools, offices, homes, churches, mosques and what have you, which are made from wood.

But the trees have many more uses than you can think of. They help us in rainfall, fertilize the soil, check erosion, are used for medical purposes, and give us oxygen to breathe and more.

Trees are so important that we must take very good care of them. We must only cut them when it is very necessary to do so. As far as possible we must plant new trees in place of those we cut and at places where there are none.

Have in mind that when the last tree dies the last man will also die. (Reportedly composed by Vida Kumi, a pupil of the local authority junior secondary school, Tinkong.)

REFERENCES

Abdulai, A. S., E. A. Gyasi, and S. K. Kufogbe with assistance of P. K. Adraki, F. Asante, M. A. Asumah, B. Z. Gandaa, B. D. Ofori, and A. S. Sumani, "Mapping of settlements in an evolving PLEC demonstration site in northern Ghana: An example in collaborative and participatory work", *PLEC News and Views*, No. 1, 1999, pp. 19–24.

Gyasi, E. A., "Development of demonstration sites in Ghana", *PLEC News and Views*, No. 18, 2001, pp. 20–28.

Gyasi, E. A. and E. Nartey, "Adding value to forest conservation by beekeeping at Sekesua-Osonson demonstration site in Ghana", *PLEC News and Views*, New Series 1, March, 2003, pp. 12–14, available from http://c3.unu.edu/plec/; also http://rspas.unu.edu.au/anthropology/plec.html.

Gyasi, E. A. and J. I. Uitto, eds, *Environment, Biodiversity, and Agricultural Change in West Africa: Perspectives from Ghana*, Tokyo: United Nations University Press, 1997.

Gyasi, E., G. T. Agyepong, E. Ardayfio-Schandorf, L. Enu-Kwesi, J. S. Nabila, and E. Owusu-Bennoah, "Production pressure and environmental change in the forest-savanna zone of southern Ghana", *Global Environmental Change*, Vol. 5, No. 4, 1995, pp. 355–366.

Owusu-Bennoah, E. and L. Enu-Kwesi, "Investigating into trees that combine effectively with field crops", *PLEC News and Views*, No. 18, 2000, pp. 20–22.

Padoch, C., "Spotting expertise in a diverse and dynamic landscape", in H. Brookfield, Christine Padoch, Helen Parsons, and Michael Stocking, eds, *Cultivating Biodiversity: Understanding, Analysing and Using Agricultural Diversity*, London: ITDG Publishing, 2002, pp. 96–104.

Padoch, C. and M. Pinedo-Vasquez, "Demonstrating PLEC: A diversity of approaches", *PLEC News and Views*, No. 13, 1999, pp. 32–34.

Pinedo-Vasquez, M., E. A. Gyasi, and K. Coffey, "PLEC demonstration activities: A review of procedures and experiences", in H. Brookfield, Christine Padoch, Helen Parsons, and Michael Stocking, eds, *Cultivating Biodiversity: Understanding, Analysing and Using Agricultural Diversity*, London: ITDG Publishing, 2002, pp. 105–125.
Zarin, D. J., G. Huijin, and L. Enu-Kwesi, "Methods for the assessment of plant species diversity in complex agricultural landscapes: Guidelines for data collection and analysis from PLEC Biodiversity Advisory Group (BAG)", *PLEC News and Views*, No. 13, 1999, pp. 3–16.

Part II

Cropping systems and related case studies

6

Management regimes in southern Ghana

Edwin A. Gyasi

Introduction

The term "management regime" refers to the techniques and methods of managing the land, water, and biota for crop and livestock production. It forms an integral part of agrodiversity, which is defined as "the many ways in which farmers use the natural diversity of the environment for production, including not only their choice of crops but also their management of land, water, and biota as a whole" (Brookfield and Padoch, 1994: 9; see also Brookfield, Stocking, and Brookfield, 1999). An understanding of agricultural management regimes is necessary as a basis for planned conservation and plant and animal diversity, and the natural biophysical resources underpinning it.

The following is a description of management regimes, including organizational practices or arrangements, as they relate to biodiversity and biophysical status in general in agricultural areas within the PLEC demonstration sites, Gyamfiase-Adenya (Map C), Sekesua-Osonson (Map D), and Amanase-Whanabenya (Map E), in southern Ghana (Map B).

The field types and their management regimes

Field types

In a sample survey of farmers and their households, the following field types or land-use stages were the most frequently encountered in all three demonstration sites:

- annual cropping
- agroforest
- house/home garden
- fallows, dominated by *C. odorata*, a notorious weed.

Virtually all the households kept livestock, most commonly goats, sheep, and the domestic fowl.

The survey showed infrequent occurrence of orchard, forest, woodlot, and edges or hedgerows. Woodlots were mostly made up almost exclusively of *Cassia siemens*. The edges encountered were in the form of cassava planted around a yam field of the farmer Kwabena Asiedu, and a line of cashew along one side of a food crop farm of the farmer Florence Akoto, all in Gyamfiase-Adenya.

Typically the agricultural holdings were fragmented into fallow and cropped plots. The plots show a mosaic pattern in Gyamfiase-Adenya and portions of Amanase-Whanabenya where land is owned communally by extended families of Akuapem people, whose forebears were migrant cocoa farmers. They exhibit a more regular pattern in Sekesua-Osonson and areas of Amanase-Whanabenya where land is owned privately by individual persons and families of Krobo and Shai/Siade people, in accord with the linear or longitudinal *huza* arrangement devised by their forebears.

The number of farms ranged from one to eight per farmer-respondent, whilst an individual farm unit rarely exceeded two hectares. Total size of agricultural land holding ranged from less than a hectare to about 300 ha per farmer.

Table 6.1 provides a summary of the farm management regimes and organizational aspects. The major advantages associated with some of them are summarized in Table 6.2, a repetition of Table 5.1.

Site preparation and tools

Sites for cropping are prepared by slashing, using the cutlass/machete, and burning the slashed vegetation. Often, trees having economic, medicinal, and ecological or some other value are left standing and even nurtured, with the food crops interplanted among them in a traditional kind of agroforestry.

The hoe is used to turn the soil and make mounds, ridges, and drainage channels. Together with the cutlass, the hoe is the tool most commonly used for sowing and clearing weeds.

By extensive burning of vegetation and the resultant destruction of faunal habitat, the indiscriminate use of fire contributes significantly towards biodiversity erosion in periodically cropped areas. Similarly, because of its damaging effects on plant propagates or seed stock in the soil, extensive usage of the hoe is suspected to be a major cause of biodiversity loss, especially in areas farmed by tenants who rely heavily on this implement. By contrast, because of its low environmental impact, the cutlass exerts a less damaging effect on biodiversity. A similar less damaging effect is achieved by *proka*, a land preparation method that avoids burning of the slashed vegetation, but rather uses it for mulch as described in Chapters 11 and 14.

Table 6.1 Farm management regimes and organizational aspects at demonstration sites

	Soil fertility regeneration/ conservation techniques				Land tenure		Tree tenure			Weed control	
Demonstration site	Three most popular	Up to three others	Food crop planting mode	Up to three principal water sources for crops	Tenancy	Non-tenancy	Land-owner	Tenant	Conservation	Up to three principal ones	Up to three others
Gyamfiase-Adenya	Bush fallow, household refuse, *proka* mulching	Crop rotation	Intermixture, sequential, line/row planting	Rain, well, stream and other surface water bodies, dew	–	–	–	–	*In situ* in farms, bush fallow, forest conservation, coppicing, taboos and reverence	By cutlass and hoe, burning, sun-drying of cleared weeds	Manual uprooting, burying
Sekesua-Osonson	Bush fallow, household refuse, *proka* mulching	Crop rotation, chemical fertilizer	Intermixture, sequential, line/row planting	Rain, well, stream and other surface water bodies, dew	–	–	–	–	*In situ* in farms, bush fallow, forest conservation coppicing, taboos and reverence	By cutlass and hoe, burning, sun-drying of cleared weeds	Manual uprooting, burying
Amanase-Whanabenya	Bush fallow, household refuse, *proka* mulching	Crop rotation	Intermixture, sequential, line/row planting	Rain, well, stream and other surface water bodies, dew	–	–	–	–	*In situ* in farms, bush fallow, forest conservation, coppicing, taboos and reverence	By cutlass and hoe, burning, sun-drying of cleared weeds	Manual uprooting, burying

Table 6.1 (cont.)

	On-farm pest control		Harvesting and storage	Labour			Gender division of labour													
							Land clearing		Land prepar-ation		Planting		Weeding		Harvesting		Transport-ation of produce to house/ market		Marketing	
Demonstration site	Principal one	Other		Self and family	Hired	*Nnoboa*	M	F	M	F	M	F	M	F	M	F	M	F	M	F
Gyamfiase-Adenya	Ash solution of *neem* tree leaves and seeds	Mixed cropping, pesticides especially for vegetables, traps, scarecrows	Manual harvesting, staggered yam harvesting through *in-situ* storage under-ground, storage in barns, under-ground burying, DDT and other agro-chemicals	1st	3rd	2nd	P	S	P	S	P	P	P	P	P	P	S	P	S	P
Sekesua-Osonson	Ash solution of *neem* tree leaves and seeds	Mixed cropping, pesticides especially	Manual harvesting, staggered yam	1st	3rd	2nd	P	S	P	S	P	P	P	P	P	P	S	P	S	P

		for vegetables, traps, scarecrows	harvesting through *in-situ* storage under-ground, storage in barns, under-ground burying, DDT and other agro-chemicals																	
Amanase-Whana-benya	Ash solution of *neem* tree leaves and seeds	Mixed cropping, pesticides especially for vegetables, traps, scarecrows	Manual harvesting, staggered yam harvesting through *in-situ* storage under-ground, storage in barns, under-ground burying, DDT and other agro-chemicals	1st	3rd	2nd	P	S	P	S	P	P	P	P	P	P	S	P	S	P

Source: PLEC field studies
P – Principally responsible; S – Subsidiarily responsible

Table 6.2 Selected management regimes/practices and their advantages in PLEC demonstration sites in southern Ghana

Practices/regime	Major advantage
Minimal tillage and controlled use of fire for vegetation clearance	Minimal disturbance of soil and biota
Mixed cropping, crop rotation, and mixed farming	Maximize soil nutrient usage; maintain crop biodiversity; spread risk of complete crop loss; enhance a diversity of food types and nutrition; favour soil regeneration
Traditional agroforestry: cultivating crops among trees left *in situ*	Conserves trees; regenerates soil fertility through biomass litter. Some trees add to productive capacity of soil by nitrogen fixation
Proka, a no-burn farming practice that involves mulching by leaving slashed vegetation to decompose *in situ*	Maintains soil fertility by conserving and stimulating microbes and by humus addition through the decomposing vegetation; conserves plant propagates including those in the soil by the avoidance of fire
Bush fallow/land rotation	A means of regenerating soil fertility and conserving plants in the wild
Usage of household refuse and manure in home gardens and compound farms	Sustains soil productivity
Use of *nyabatso*, *Neubouldia laevis*, as live stake for yams	The basically vertical rooting system of *nyabatso* favours expansion of yam tubers, while the canopy provides shade and the leaf litter mulch and humus. It, also, is suspected that *nyabatso* fixes nitrogen
Staggered harvesting of crops	Ensures food availability over the long haul
Storage of crops, notably yams, *in situ* in the soil for future harvesting	Enhances food security and secures seed stock
Conservation of forest in the backyard	Conserves forest species; source of medicinal plants at short notice; favours apiculture, snail farming, and shade-loving crops such as yams

Source: PLEC fieldwork since 1994

Cropping patterns

The dominant crop is cassava, followed by maize – both annual or seasonal food crops. The cassava shows a wide variety of types, notably the following, identified by their popular local names: *ankra*, *agbelitomo*, *katawire*, *abontem*, *asramnsia*, *tuaka*, *biafra* or *agege*, *trainwusiw*, *bankye nsantom*, *gbezey*, and *agbeliatsilakpa*. The last two are recent introductions from Benin and Togo, which underscores the importance of cross-border movements in agrodiversity. Other important annual/seasonal crops include cocoyam and assorted yams. The widest variety of yam appears to be in Sekesua-Osonson. It is reflected by the diversity of yam names in the local Dangbe-Krobo language, notably the following:

- *alamoa*
- *alamoa kpeti*
- *alamoa gaga*
- *kpanya alomoa*
- *nyamatso*
- *kyramakyira*
- *alamoayiblitse*
- *odorno*
- *tsom kani*
- *kani*
- *hier*
- *kofokani*
- *klowalamoa*
- *alamoakpoto.*

In Gyamfiase-Adenya, the species of yam identified by their local names include the following:

- *apuka*
- *nkani*
- *ode*
- *nkuku*
- *labreko*
- *ntetareso*
- *mensa*
- *kokoo ase bayere*
- *nkamfo*
- *afasew*
- *basajor*
- *oboobikwao.*

The next two chapters discuss yams in greater detail.

Vegetables and legumes, notably pepper, tomato, okro, garden eggs, and beans featured most prominently in Sekesua-Osonson. Among the perennials, plantain, and the oilpalm was the most common. Others include citrus, cocoa, and coconut. The rich diversity of crops helps to secure food by minimizing or preventing complete crop failure. Mixed cropping, a traditional practice of securing food supplies, still remains the rule. It is dominated by cassava and underlies the rich crop biodiversity, as does the practice of using breaks in canopies in cocoa and oilpalm orchards for food crops and vegetables. However, monoculture, centred especially on maize, is gaining in popularity, as is row or line planting. This trend is reinforced by the modern practice of chemical fertilizer application promoted by official government policy. Sequential cropping is common. However, crop rotation did not appear very widespread. Mainly perennials are planted in February–March, annuals in March–May and August–October, and vegetables in January–March and

May–September. These temporal variations in planting, like the harvesting of perennials in May–August and November–December, of annual crops in June–September and December, and of vegetables in December, are in accord with the bimodal rainfall pattern and two intervening dry seasons. The staggered planting and harvesting minimizes the risk of food shortage within the year.

Materials saved from a farmer's previous harvest were the most common source of seeds and other planting materials. Other sources were:

- market centres, including the local periodic market
- agricultural extension agency
- Adventist Development and Relief Agency (ADRA), which supplies grafted mangoes in Sekesua-Osonson
- PLEC, which supplies citrus and germinated oilpalm seeds.

Soil fertility regeneration

The quality of soils is pivotal to sustainable cropping. In the demonstration sites, the most popular means of soil conservation or fertility regeneration is through bush fallow or land rotation, whereby after a period of cropping lasting up to about two years, the land is left to fallow for an average of one to four years. The fallow period allows the soil to regenerate by resting and by the litter of naturally sprouting plants, especially in long fallows. It also enhances natural floral diversity through the regenerating vegetation. Especially the Krobo people in Sekesua-Osonson use household refuse to enrich soils in home gardens.

Various forms of mulching are practised. They include the use of cleared weeds and residue of harvested crops, and the *proka* system whereby the vegetation cleared in the course of land preparation is left in place, without burning, and the crops subsequently planted within it so that they may benefit from the moisture conserved and humus generated by the vegetation mulch.

Other soil improvement practices include crop rotation and the use of chemical fertilizer, which is severely constrained by poverty and the poor supply situation. A PLEC attempt to encourage soil conservation through demonstration of the use of stone lining to check erosion in sloping areas in Gyamfiase-Adenya and Sekesua-Osonson met only a mixed reaction, apparently because of the arduous nature of the work involved.

Weed and on-farm pest control

Weeds are a major problem in farms. Mainly they are controlled by weeding by cutlass and hoe. Sometimes they are uprooted manually by hand. Subsequently they may be left as mulch, or to sun-dry for burning. Weeds that do not dry and burn easily may be buried.

Insect pests are often controlled by ash from burnt biomass, and occasionally by a solution concocted by farmers from leaves and seeds of the *neem* tree, *Azadirachta indica*. Modern commercial pesticides are used sporadically, especially for vegetables. Farmers also claim that intermixture of certain crops is able to repel some insect pests. The most common method of controlling rodent pests is by trap, of which there are various kinds. Scarecrows are used to scare off birds that feed on unharvested crops, especially maize.

Water sources and management

Crop farming is rain-fed. Farmers do, occasionally, use water from nearby streams for their vegetable farms during dry spells. Rain, streams, wells, boreholes, and lakes are the sources of water for livestock and domestic use.

Often, moist or seasonally flooded depressions are used for water-loving crops such as sugarcane all the time, and for vegetables during dry periods. Some farmers make drainage channels to drain excess water in farms. A few use stone lining to trap rainwater and to minimize its erosive impact on soils. A common water conservation practice is mulching, including *proka*. The use of mounds, especially for yams, is yet another moisture conservation practice.

Harvesting and storage

In all cases, harvesting is done manually by hand, with or without the aid of simple implements, notably the machete/cutlass and the hoe. The maize cob is pulled off the stalk by hand. Stubborn ones are cut off by cutlass. Peppers, tomatoes, garden eggs, okro, and virtually all vegetables and condiments are plucked by hand. The palm fruit is harvested by machete. The harvester does this whiles standing on the ground if the palm tree is short. If the tree is tall, it is climbed or an improvised platform is used for the purpose of harvesting. The cocoa pod is harvested by cutting it off from the tree using the machete or by simply squeezing it off by hand if the pod is within easy reach. Pods occurring at higher levels are plucked by a tall metal-tipped wooden pole. Cassava and cocoyams are harvested by holding the stem to pull out the tubers. A cutlass is used to dig out tubers stuck in the ground. However, in cases where the yam has penetrated deeply into the ground, the tool most commonly used for harvesting is a metal-tipped wooden pole.

Two aspects of the manner of harvesting yams are particularly favourable for food security and biodiversity conservation. One is the staggering of the harvesting over a long period. In this practice, only one or two matured tubers on a stem is/are harvested at a time, with the rest left *in situ* in the ground to be harvested later. A second aspect is the practice of leaving unharvested small yam nodules in the ground for them to regerminate in a process that may continue for years. These and other aspects of yam management, including the practice of live staking, are particularly well developed in home garden agroforestry systems in Sekesua-Osonson and other migrant Krobo areas as discussed below.

Regarding storage, a popular practice is the use of wooden barns in the farm or at home, particularly for maize and cassava. Often freshly uprooted cassava tubers and, sometimes, those of cocoyam are buried in the ground for storage.

Water may be sprinkled on the burial ground to maintain a proper moisture-temperature balance during spells of dry weather. Peppers are commonly stored in sacks or on the bare floor after sun-drying.

Farming systems

The practices outlined above are embodied in various farming systems. Foremost among the systems are essentially traditional biodiverse bush fallow and home garden agroforestry (Table 6.2).

Bush fallow

The bush fallow system proceeds on a rotational basis in plots around fixed settlements.Through any of the following practices, it conserves biodiversity:

- controlled use of fire to clear vegetation on a selective basis
- use of non-burnt vegetation for mulching
- minimal tillage
- use of environmentally low-impact tools
- intercropping, often among trees left *in situ*.

A factor in the high crop diversity that often characterizes the system is usage of varied edaphic conditions, breaks in canopies, and other ecological niches within the farm. Table 6.2 summarizes the advantages of this and other management regimes.

Home garden and non-home garden agroforestry

As discussed in some detail in Chapter 12, home garden agroforestry, i.e. an agroforestry unit located adjacent to the home, is particularly well developed in areas settled by migrant Krobo farmers and other Adangbe-speaking people, notably Sekesua-Osonson.

Like the type of agroforestry found away from the home, home garden agroforestry is vegecultural in character. Both types centre on African species of root and tuber crops, notably yam, *Dioscorea*. However, home garden agroforestry contains a wider diversity of other crops. Although cocoa has generally diminished in importance in migrant Krobo areas, it still shows higher concentrations in the home garden agroforestries. Other crops still showing greater concentrations in home garden agroforestries include the following:

- peppers, other condiments, and certain leafy vegetables in regular demand by the kitchen
- plantains and bananas
- fruit trees such as mango, avocado, citrus, and *Chrysophyllum albidum* (called *adesaa* by Twi-speaking Akan people, and *alatsa* by Ga-Adangbe-speaking people, including the Krobo).

Trees among which crops may be interplanted include the following:

- those difficult to fell
- those perceived as sacred

- those having medicinal value
- those performing unique ecological functions
- those having economic value, notably oilpalms, kola, and timber species.

Yam management within agroforestry and other land-use systems, most especially by Krobo farmers, is characterized by the following practices that conserve soil and agrodiversity, makes for food security, or achieves all:

- minimal till, a method that avoids mounds by sowing directly in a hole drilled in the soil with little disturbance
- before sowing, dressing of the yam hole by decomposing leaves, cocoa husk, and other biomass to enrich soils
- live staking that makes use of various plants, particularly *Neubouldia laevis*, called *nyabatso* by Adangbe-speaking people, and *osensrema* by Twi-speaking Akuapem people
- staggered harvesting of the diversity of yams
- post-harvest *in situ* storage of the smaller yam tubers right in the soil as seed stock.

Aspects of these are elaborated upon in Chapter 7.

Like the non-home garden agroforestry, home garden agroforestry makes for food security and a sound agro-ecosystem by any or all of the following:

- an environmentally low-impact tillage using the cutlass, hoe, and fire on a parsimonious basis, i.e. if it is applied at all
- minimizing soil erosion through a canopy of vegetation
- making more effective use of the different soil nutrients by the different demands exerted by the different crops
- maintaining ecological stability by the combination of crops with trees
- spread of the risk of complete crop failure among the different crops
- making for a more balanced diet through the diversity of crops.

Explanation of the development of home garden agroforestry in migrant Kroboland lies in the *huza* arrangement whereby family houses are constructed linearly along a common base. From the base, farming proceeds in the same general longitudinal direction uninhabited by other dwelling units. This is in contrast to the situation in areas characterized by nucleated homes, which crowds out gardens. In most homes in *huza* areas crops are sustained, at least partially, by household refuse. In the valleys where many of the homesteads are located, an additional source of plant nutrients is soils washed from the uphill (Gyasi, 2002).

Animal husbandry

Very nearly all farmers keep domesticated animals in addition to cultivation of crops, even though these two activities were hardly integrated to mixed farming, with the major exception of:

- the use of crop residues, most notably those of cassava and plantain, to feed animals
- the occasional tethering of sheep and goats in nearby fallow fields to forage
- the sweeping of animal droppings to fertilize home gardens, a practice which, more often than not, is carried out with indifference.

The principal livestock are chicken, goats, and sheep. Ducks, rabbits, and cattle were kept on a very limited basis. On the whole, animals were managed on a free-range basis, even though goats and sheep are often tethered during the day to prevent damage to nearby farms. At night they are commonly confined in the household compound. Fowls may perch on neighbouring trees. In Gyamfiase-Adenya seven farmers confined and fed their sheep and goats in pens permanently.

Gender division of labour

Mainly the owner and his/her family operate farms. Other forms of labour include that hired, especially for land clearing and weeding. Also common is the *nnoboa* system whereby farmers work in each other's farms on a reciprocal basis.

There is considerable division of labour. Mainly males clear land, while its preparation is mainly by females, as is transportation of produce to home and the market. However, the tasks of planting, weeding, and harvesting are shared more or less equally among the males and females.

Resource tenure

Land tenure

As elaborated upon in Chapter 18, in the demonstration sites, as in Ghana generally, land is owned as common, group, clan, or family property. The land is freely used by those owning it, but typically paid for in cash or in kind by others.

Thus, there are two principal means of access to land. The first, the most fundamental, is based on kinship. It involves no payment, unlike the second, which involves payment based on a tenancy contract. Access to land may also be achieved by purchase. Such purchases are private property or, eventually, they become ancestral or extended family property. Many of the farmers operate as owner-occupiers on the basis of kinship. However, others operate as tenants.

The land-tenure situation is a complex one. Aspects of it, notably the following, may help to explain biophysical status including the status of biodiversity in agricultural areas:

- security of the tenure
- the spatial layout of the landholdings and the attendant settlement pattern.

Tree tenure

Tenure with respect to trees is just as complex. By traditional customary law or convention, timber trees are owned exclusively by the landowner. However, under modern Western-type law, which supersedes the customary, ownership of such trees came to be vested in the modern state or government. Presently this arrangement is undergoing review.

Tenants may have free access to other tree species, notably firewood for subsistence, but without compromising the ultimate ownership right of the landowner. Certain species of naturally occurring fruit trees are regarded as a common property freely accessible to all for basic subsistence. The ownership status of trees planted by a tenant remains ill-understood and, therefore, requires further investigation. How the system of resource access and distribution relates to the use of land (with special reference to biodiversity and biophysical status) is the subject of Chapter 18.

Conservation

For ecological, agronomic, economic, and other utilitarian purposes, commonly trees are conserved in farms and fallow areas as part of the agricultural system. For similar purposes, patches of forest may be kept in reserve. Sometimes, such forests are perceived as sacred, as is the case with the Gyamfiase forest grove. Coppicing is a fairly widespread conservation practice, especially among Krobo people. The tree most popularly coppiced for firewood and charcoal is *Cassia siemens* (Plate 8). Trees may be protected by taboos.

Quantitative analysis and conclusion

Quantitative analysis

Among the factors that may explain the variations in agrobiodiversity in the demonstration sites are the following management aspects:

- farming distance from a farmer's house
- use of household refuse
- extent and manner of use of fire/burning
- use of the *proka* mulching and no-burn system
- use of chemical fertilizer
- frequency of use of the hoe
- use of ecological niches, e.g. breaks in farm canopy, moist patches, and termite mounds
- available household labour: size of farm labour originating from household
- whether or not farming is carried out on a tenancy basis
- security of tenure
- systems of harvesting and storage
- share-cropping or tenancy arrangement.

Towards a quantitative analysis of the relationships, a matrix might be developed from PLEC survey data using Table 6.3 as a framework. In the table, the first

Table 6.3 Matrix for quantitative analysis of variations in biodiversity

Field no.	Dependent variable	Independent variables							Land-use/field type
	Species diversity/richness (of field)	Distance	Tenure					Soil richness	
			Tenancy						
			Owner operated/ owner held	Share-crop	Cash payment	Exclusive right to firewood (Y/N)	Lease length		
1.									
2.									
3.									
4.									
5.									
6.									
7.									
n.									

Table 6.3 (cont.)

Field no.	Independent variable														
	Topographical features and ecological niches					Management: frequency of use of					Household characteristics				
	Streams and ponds	Depressions	Termite mounds	Stones	Other	House-hold refuse/ farm-yard manure	Chemical fertilizer	*Proka* mulching and no-burn	Hoe	Other	Size	Gender M F	Age	Net migra-tion	etc.
1.															
2.															
3.															
4.															
5.															
6.															
7.															
n.															

column shows the numbering of the fields, which were surveyed according to households. The second column shows the corresponding land-use types as indicated below:

1. annual cropping
2. agroforestry: non-home garden type
3. fallow
4. orchard
5. native forest
6. home garden: agroforestry type
7. home garden: non-agroforestry type
8. livestock
9. other.

They are a modification of the categories recommended by PLEC's Biodiversity Advisory Group (Zarin, Huijun, and Enu-Kwesi, 1999)

The third column of Table 6.3, shows richness of plant species found in each field, which is designated the dependent (Y) variable. The remaining columns contain possible explanatory variables (X1, X2,..........Xn).

Conclusion

Management regimes and organizational forms are diverse in the demonstration sites. Variations in richness of the biota appear to be related to this diversity. The challenge is to give quantitative expression to the relationship.

REFERENCES

Brookfield, H. and C. Padoch, "Appreciating agrodiversity: A look at the dynamics and diversity of indigenous farming systems", *Environment*, Vol. 36, No. 5, 1994, pp. 6–11, 36–45.

Brookfield, H., M. Stocking, and M. Brookfield, "Guidelines on agrodiversity assessment in demonstration site areas" (Revised to form a companion paper to the BAG guidelines), *PLEC News and Views*, No. 13, 1999, pp. 3–15.

Gyasi, E. A., "Traditional forms of conserving biodiversity within agriculture: Their changing character in Ghana", in H. Brookfield, C. Padoch, H. Parsons, and M. Stocking, eds, *Cultivating Biodiversity: Understanding, Analysing and Using Agricultural Diversity*, London: ITDG Publishing, 2002, pp. 245–255.

Zarin, D. J., G. Huijun, and L. Enu-Kwesi, "Methods for the assessment of plant species diversity in complex agricultural landscapes: Guidelines for data collection and analysis from the PLEC Biodiversity Advisory Group (PLEC-BAG)", *PLEC News and Views*, No. 13, 1999, pp. 3–15.

7

Yams: Traditional ways of managing their diversity for food security in southern Ghana

Essie T. Blay

Introduction

Yams (*Dioscorea* spp.) occur throughout the tropics. At least seven wild species exist in West Africa but six main types are cultivated. The cultivated ones are white yam or eight-month yam (*D. rotundata*), yellow yam or 12-month yam (*D. cayenensis*), greater or water yam (*D. alata*), lesser or Chinese yam (*D. esculenta*), bulbil-bearing yam or air potato (*D. bulbifera*), and trifoliate, cluster, or bitter yam (*D. dumetorum*) (Irvine, 1979; Degras, 1993; Blay, 2002; Kranjac-Berisavljevic and Gandaa, 2002; see also Chapters 8 and 17; Plates 9–11).

The transitional and northern savanna regions constitute the centre for commercial yam production in Ghana (see Chapter 8; Map B). The cultivation of yams in Ghana is concentrated around Northern, Ashanti, Brong Ahafo, Volta, and Upper West regions (Abbiw, 1990). However, the crop is grown throughout the country both as a commercial and a subsistence crop. The tubers are a rich source of vitamin C, potassium, phosphorus, and protein (4–8 per cent of fresh weight). Yam is regarded as a prestigious food in many communities in Ghana. The tubers are served boiled, fried, baked, or roasted, or prepared into porridge similar to Irish stew. Certain yam dishes, notably *ɔtc* (Akan-Twi word for a mashed yam preparation), are used in several local customs and celebrations.

Major festivals are woven around yam harvest. In some of the major yam-growing areas in Ghana, prescribed rites must be performed at the beginning of the harvest season before the tubers are released for consumption. Yam is also an important food security crop. Some yam varieties have a very long shelf life compared to the other non-cereal starchy staples. As such, many rural households in many yam-growing areas in southern Ghana, especially the forest-savanna transitional areas, maintain a rich array of yam for their year-round sustenance and income generation. To ensure the availability of and maintain the diversity of yams, the farmers have evolved very effective traditional production and management practices. The bush yam (*D. praehensilis*), for instance, is sometimes grown in one spot for several decades and managed as a perennial crop.

This chapter discusses the distribution of yams in the different land-use types, indigenous agronomic practices adopted by the peasant farmers, and the strategies for conservation and maintenance of diversity for food security with reference to PLEC demonstration sites in southern Ghana (Map B).

The diversity of yams

Yams cultivated in southern Ghana belong to six major species. Each species is represented by many land varieties, some with several strains. Some species, for instance white yams (*D. rotundata*), are used in the preparation of a wide range of dishes, while others, such as trifoliate yams, have limited use.

The types of yams grown by farmers are determined by dietary preferences, the requirements for the preparation of specific dishes for customary, religious, and other celebrations, and the market value. The white yam is used more extensively in customary rites and celebrations, while the water yam enjoys the greatest longevity with a consequent high market value during the lean season. Other factors that come into play in the choice of yam cultivars on farmers' holdings include palatability. The Akyem/Akim people reportedly better appreciate the trifoliate yam than the Krobo people.

Another basic determinant is the availability of planting material. The major yam planting material is the edible tuber. This places a severe restriction on the availability of planting material. The problem is further compounded by the fact that a significant proportion of the less commercially important yam varieties are heirloom varieties, with restricted distribution.

Agronomic characteristics such as ease of cultivation and conservation, amenability of the variety to management as a perennial, and shelf life of harvested tubers are also considered by farmers in selecting the varieties to cultivate. In a few instances, the driving force is simply an interest in maintaining heirloom varieties for posterity.

Water yams (D. alata)

The most commonly cultivated yams in southern Ghana are the water yams. Water yam is a vigorous-growing yam with relatively large thick leaves. It is distinguished from the other species by its pink or green angular, winged stem. This yam is very popular among farmers owing to its long shelf life of over one year in some cultivars, the ease of cultivation, and the commercial value. Tuber cuttings from water yams sprout easily; even the peel exhibits sprouting potential. In addition to the tubers, some water yam varieties produce aerial bulbils that are also used as "seed". Different varieties vary in texture, taste, and keeping qualities; hence farmers usually plant several varieties to cater for their needs. Up to 10 different varieties may be grown by a single farmer.

Bush yam (D. praehensilis; Plate 9) and cultivated yam (D. rotundata)

The second most popular yam type in terms of frequency of occurrence is the bush yam (wild progenitor of *D. rotundata*), locally known as *Kookoose bayere* (Twi language) and cultivated white yam (*D. rotundata*). These are grown under permanent trees, mainly cocoa, in the cocoa-growing areas, hence the name. The bush yams have rounded stem with varying pigmentations and various levels of thorniness, often bearing cataphylls. Owing to their occurrence under the canopy of tall trees, they are very vigorous climbers with very long internodes. In the Krobo area, the *Newbouldia* sp., locally known as *nyabatso*, with its deep, soft tap-root system and low canopy shade is commonly used as a live stake for the yams. The bush yam (*Kookoose bayere*) derives its popularity from the ease of cultivation, and the fact that it can be managed as a perennial. The tubers are frequently left in the ground and harvested piecemeal as needed. Tubers left unharvested in the mounds remain palatable even after dormancy is broken, until leaves are fully developed. New plants developed from such tubers produce multiple yields with time. Bush yam tubers do not store well and must be consumed within a couple of weeks after harvest. This makes them relatively unimportant commercially. The comparative advantage of bush yams is the ease of perenniation and, hence, its food security value.

Between 78 and 92 per cent of farmers in southern Ghana cultivate four to 10 different varieties of bush yams.

Yellow yam (D. cayenensis)

Of the remaining four species, the yellow yam or *nkani* (Akan-Twi name) is the most important. This species has a round, often smooth stem and produces a characteristic yellow-pigmented tuber that is revered by members of the Akuapem tribe. Unlike the trifoliate yam, the yellow yam tuber is not bitter and does not normally form tuber clusters. It is also easily distinguishable from the trifoliate yam by its cordate leaves and glaborous stems. Four cultivars of yellow yams were encountered in the southern Ghana demonstration sites.

Bitter yam (D. dumetorum; Plate 10)

The bitter or trifoliate yams are locally known as *nkamfo* or *akim baale*. Unlike the yellow yams, these are better known at Amanase-Whanabenya and Gyamfiase-Adenya than at Sekesua-Osonson demonstration site (Maps B, C, D, E). Farmers in Amanase and Gyamfiase cultivate one or two varieties of the bitter yam. It is a vigorous yam with round, hairy and/or thorny stems. The tubers vary in pigmentation from white through cream to various shades of yellow.

Aerial yam (D. bulbifera; Plate 11) and Chinese yam (D. esculenta)

The aerial yams and Chinese yams occur sporadically and are of limited importance. The Chinese yams are commonly referred to as "potato". The species has thin, glabrous, less vigorous vines, with thin, non-waxy, rounded leaves. It produces a cluster of many small oval tubers, up to 10 cm long, with very thin skin. Only one variety of the Chinese yam, *obo aduanan* (Akan-Twi expression), is found in the demonstration sites. The Chinese yam is one of the most nutritious yams. However, it is relatively little known with equally limited distribution and commercial importance

In the case of aerial yams (*D. bulbifera*) the aerial bulbils are the edible part. The small, relatively hard underground tuber is exclusively used as a planting material. This yam is not popular because closely related, poisonous, wild types exist in Ghana. Therefore only those who are able to distinguish the true *D. bulbifera* consume it.

Distribution of the different yam types in the various land-use systems

The various yam species are grown in different land-use systems (Figure 7.1). The bush yams are grown predominantly in tree-associated land-use systems, while the yellow yams and water yams are mainly limited to the annual cropping systems with no tree canopy.

The bitter or trifoliate yams occur in all the four land-use systems where yams are cultivated, but they are mainly grown in home garden agroforestry and annual cropping systems. The aerial yams (*D. bulbifera*) and the Chinese yams (*D. esculenta*) occur infrequently in all land-use systems with the highest diversity occurring in home garden agroforestry and annual cropping systems. Generally, the widest diversity in yams is found in home gardens, where their maintenance is conveniently handled by the farm family.

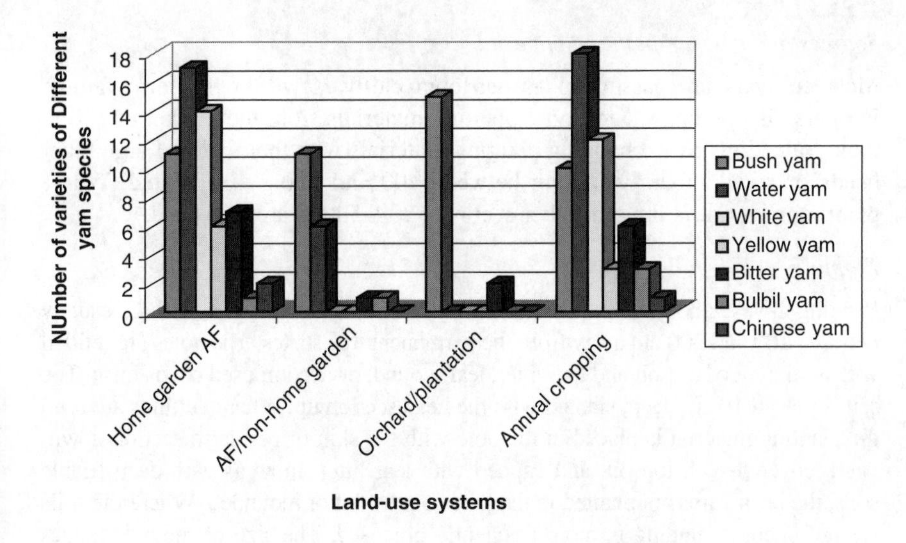

Figure 7.1 Diversity of yams in different land-use systems.

Indigenous cultural practices adopted for the different types of yam

Bush yams (D. praehensilis) or wild rotundata

Ecology

Bush yams are grown in the home garden agroforestry systems, agroforestry non-home gardens, orchards and plantations, and in annual cropping systems, but their cultivation is mostly associated with orchards and plantations.

Time of planting

New bush yam plantings are normally made between December and February but they may be planted in March. During the minor season, yam heads obtained from tubers, which are milked for consumption in July, may also be replanted. Such planting materials remain dormant in the planting holes until about February when dormancy is broken. Subsequently they sprout at the same time as the major-season plantings.

Land preparation

On fallows or new farmlands, the bushy undergrowths are slashed at the time of planting.

Sources and types of planting material

Most cultivars are passed on as heirloom cultivars within the farm family. Farmers also purchase additional planting materials from the market or add to their collections by exchanging planting materials with their peers. Large tuber heads or small tubers weighing between 400 and 500 g are planted. Where planting material is limited, tuber sections, 200–500 g, are also used.

Planting

Planting holes, about 30–50 cm in diameter and 40 cm deep, are dug, usually between 100 and 200 cm away from the permanent tree stakes. The holes are refilled with a mixture of topsoil and leaf litter, leaf mound, or decomposed cocoa husk. The sett is placed 10 cm deep, slanted with the head section up. Where cuttings are used the planting material is placed in the hole with the skin or periderm section downward, covered with topsoil, and capped with leaf litter. In areas with deep friable soils, the bush yams are planted in the holes on the flat or mounded. Where the soils are not deep, mounding is more frequently practised. The size of mounds ranges between 50 and 100 cm in diameter and between 15 and 20 cm in height. Mounding is frequently done after the yam has started sprouting.

Staking

Bush yams are exclusively grown on stakes. Plantation crops such as oilpalm, or cocoa serve as live stakes. However, certain trees with deep tap-root systems, fewer lateral roots, or which root easily are commonly used as stakes. These include *Newbouldia laevis*, *Zanthoxylum xanthozyloides*, *Horrlarhena floribunda*, *Nesogordonia papaverfera*, *Ficus exasperata*, *Spondias mombin*, *and Mallotus oppositfolium*. Farmers manage the live stakes by root pruning, coppicing, or pruning off some branches at the beginning of the planting season. Young vines, about 60–100 cm long, are trained to leader stakes that are placed 20–30 cm from the planting hole and leaned on to the permanent stake. The leader stakes may be bamboos, rachis of palm fronds, sticks, water sprouts from trees, or simply old vines from previous years' growth that serve as trellises.

Harvesting

Bush yams are ready for harvest between July and September (seven to eight months after planting). Milking or pricking usually commences at the end of July. It involves careful removal of the soil from the mound or planting hole to expose the tubers. The ware yam is detached with the aid of a cutlass leaving only the corky section of the head (corm) and its propping roots intact. The corky head may be covered immediately or left uncovered until fresh tubers are initiated from the corm (Plate 9).

Full-term tubers are harvested after nine months. To harvest the full-term tuber, the soil is carefully loosened a little at a time with the aid of a cutlass. The loose soil is removed by hand to expose the top section of the tuber. The process is continued until the entire tuber is freed (Plate 5). Where tubers are very long or buried deep, a chisel is used for digging the soil around the tubers. In the case of extremely long yams, the farmer may dig a large and deep hole near the tuber being harvested and stand in the hole in order to reach the tuber more easily without straining the back. To harvest very long tubers, an alternative strategy adopted by farmers is to first remove the soil around the head of the tuber. A branch from a nearby tree is then bent down and the tuber is tied to the branch with the aid of a twine. The hole around the tuber is then filled with water overnight. The water softens the soil around the tuber, and the force in the bent branch then pulls the tuber out.

Tuber sizes

Tubers vary between 30 cm and over 50 cm in length. Some varieties are very long, measuring over 90–120 cm, e.g. the variety *aku*.

Storage

The bush yams are usually kept in the field, and tubers are detached from the plants as required. Harvested tubers have a shelf life of only a few days, after which the cooking quality deteriorates rapidly. They are therefore kept for only one to three days to a maximum of one week. One variety, *kookooase fitaa* is reported to be storable for six months to one year.

Uses

Bush yams are valued for their palatability and food security value. Some cultivars are also used for special celebrations. They are predominantly used in the preparation of the *ampesi* dish (boiled starchy staple served with stews or sauce) or pounded in *fufu*.

Maintenance strategies

Farmers maintain the crop by *in-situ* conservation. The harvesting is staggered. Tubers are harvested from the standing crop as needed during the harvesting season. The tubers are sometimes left in the holes to sprout and produce new growth during the growing season. Subsequently they produce even heavier yields. In addition to managing the plants as perennials, fresh plantings are made yearly. Also, farmers maintain sexual seeds that drop, sprout, and produce many volunteer plants.

Greater or water yam (D. alata)

Ecology

Greater or water yams do better under full exposure to sunlight. The greater yams are normally grown in home garden agroforestry or annual cropping systems. They also occur in a very few agroforestry non-home garden land-use systems. Water yams are grown on freshly cleared or already farmed lands. They are usually, but not always, planted as the first crop in a mixed cropping system. The planting order adopted depends on availability of planting material.

Sources of planting material

Most farmers, especially newcomers, obtain planting materials of water yam from the market. Portions of exceptionally tasty tubers that are purchased for home consumption are also saved for planting. Generally, farmers save the heads of ware yams that are consumed by the farm family for replanting. A number of farmers plant heirloom varieties. Farmer-to-farmer exchange of germplasm also plays an important part in planting material supply.

Planting material and planting

Tuber heads or cuttings from ware yams are used as planting materials. Small tubers generated after milking are also used as planting material. The sizes of the yam setts vary from 200 to 250 g, but larger pieces weighing 500 g are sometimes used. Tuber yield is usually proportional to the size of planting material. However, in some varieties such as "matches", very small cuttings, the size of a small matchbox, produce very large ware yams. Water yams generally sprout very easily. Even the peeled skins of some varieties are credited with sprouting ability. They are planted in February in areas with early rains, but are mostly planted between March and April.

Nursery

The "sett" may be direct sown or nursed and transplanted. Where nursing is practised, the "setts" are nursed on the flat, either in a single layer, or up to three layers, and covered with dead leaves, soil, or a mixture of the two. Nursing lasts for two to four weeks, but seedlings may be raised in the nursery for up to two months before field transplanting.

Field planting

The yams are planted in holes on the flat, in large mounds with no planting holes under them, or in holes which are mounded over. The planting holes range from 20 to 60 cm in diameter and are 30 to 60 cm deep. Mound sizes vary from 50 to 85 cm wide and 10 to 25 cm high. The setts are planted at 10–25 cm deep depending upon the depth of the planting hole. In areas with less friable soils the planting holes are mounded over. After planting, the holes or mounds are capped with dead leaves. Very vigorous yam varieties such as *alamoa poto* are usually planted in a hole without mounding.

Staking

Water yams are staked. Trees left *in situ* or deliberately planted serve as stakes. Up to eight or more mounds are constructed around a single large tree. The desired attributes of trees used as live stakes include an open canopy, many branches, deep tap-root system, leguminous, wind resistance, and evergreen growth habit. Trees with straight stems and a compact canopy are also used. The live stakes are pruned leaving the optimum canopy and number of branches for effective support of the yam vines. Some trees are also coppiced as a way of managing the shade imposed on the yams. Where the live stake is located far from the yam mounds, leader stakes are used to link the vines to the live stakes. During staking, emerged vines are trained on the leader stakes when they are 50 cm to over 120 cm long with a mode of 60–100 cm. Staking may be done earlier when vines are around 30 cm long or later at over 120 cm long.

Harvesting

Water yams produce edible tubers five to six months after planting. Some tubers are harvested between six and eight months after planting. The bulk of the crop is harvested at between nine and 12 months after planting. In a few instances, harvesting may be delayed a little beyond 12 months when all the leaves are completely dried. Harvesting procedures are similar to those of the bush yam. Some water yams tend to have more shallow tubers making them easier to harvest. In some varieties such as *afasew nanka* and *alamoa poto* the tubers tend to pop out of the soil as they mature, making them easy to harvest. The tubers of these are continuously earthed up to prevent their exposure until harvesting. This practice preserves the cooking quality. The length of harvested tubers varies between 10 and 40 cm. Some cultivars have tubers exceeding 40 cm in length. For instance, *alamoa gaga* and *osoaba* have tubers measuring 90–118 cm in length.

Storage

Tubers may be harvested from the field as required. However, most water yams are harvested once. The tubers are stored in barns, on platforms, or stored covered under trees. Water yams store very well and may be kept in storage for over one year. The palatability of the tubers improves with storage as the tubers become less watery.

Maintenance

Farmers maintain their water yam germplasm by yearly planting. Also purchases of planting materials from the market ensure that the cultivars at a risk of extinction are replaced. The plants also regenerate from pieces of tubers left in the field after harvest. In bulbils-bearing or seed-producing varieties, the bulbils or seeds that drop from the vines produce volunteer plants and serve as organs of perenniation.

Special attributes and uses

Water yams are known for their ease of sprouting. They are cherished for their relatively long shelf life, palatability, and food security value. A few farmers also generate income from them. Some cultivars such as *afasew bore* are used for special festivals. The yields of water yams are exceptionally high. The tubers are commonly boiled for *ampesi* or made into porridge. Some cultivars are also suitable for roasting.

White yam (D. rotundata)

Ecology

The white yams are predominantly grown in home garden agroforestry systems. They are also cultivated as part of annual cropping and agroforestry non-home garden systems. The crop is grown between February and March, but may be planted as early as January or late in April.

Land preparation

White yams are planted on freshly cleared or on previously cultivated lands as a first crop in a mixed cropping system or after other crops. The planting material is grown in holes on the flat, in mounds, or in holes over which mounds are made.

Planting

Heads, cuttings, or small whole tubers are used. Sett size ranges from 250 to 500 g, but as much as 1–2 kg setts may be used. The depth of the planting hole ranges between 30 and 50 cm with a diameter of 30 cm to over 50 cm. The sett is placed 10–30 cm deep in the planting hole depending upon the sett size and the vigour of the cultivars. Mound sizes ranging from 10 × 55 cm to 15 cm × 55 cm are formed over the planting holes depending upon soil structure and the vigour of the variety. No special order is followed in the planting of white yam in a mixed cropping system. In a few cases shallow-rooted, compact crops such as groundnuts or some vegetables like shallots are planted on the mounds. The planting material is sometimes nursed for one to four weeks prior to field planting.

Staking

The staking technique is as outlined for the bush yam and water yam. The distance from the live tree stake to the base of the leader ranges from 70 cm to over 120 cm. The vines are trained along the leader stakes when they are 60–100 cm long. Vine length may sometimes reach 150 cm before staking.

Harvesting and storage

Tubers mature five to six months after planting but are harvested at seven to eight months after planting. Tuber sizes may vary between 30 cm to over 40 cm. The harvested tubers are stored in a hole covered with leaves or soil, tied to poles, or stored in barns for a period of four to 12 months.

Maintenance

White yam is mainly maintained by yearly planting, but may also be retained in the field as in the bush yam.

Special attributes and uses

White yams are palatable and are used for festivals such as the fetish yam festival. They are also important for food security. The tubers are used in the preparation of fried yam chips, *ampesi*, *fufu*, mashed yam, balls, and porridge.

Yellow yam (D. cayanensis)

Ecology

The yellow yam is grown in annual cropping systems. Field planting is normally done between February and March, but it is sometimes planted as early as January or late in April.

Land preparation and planting

Field planting is done on freshly cleared land or previously cropped fields. Usually all tree cover is removed. Yellow yams are predominantly grown as a first crop. Yam heads, cuttings or small whole tubers weighing 200–500 g are used for crop establishment. The planting hole is dug to a depth of 30–60 cm and a width of 30–60 cm. The setts are planted in holes, mounds, or both. Depth of planting is 10–30 cm. Mound sizes are relatively large, varying from 50 × 15 cm to 122 × 30 cm. The setts may or may not be nursed. Nursery practices are the same as for the other yams.

Staking

Mostly, non-live stakes are used. Where live stakes are used, they are managed by coppicing. Leaders are also used to facilitate staking. The distance from the planting hole or mound to the live stake varies between 120 and 145 cm.

Harvesting and storage

The yellow yams produce tubers as early as four months after planting, but harvesting may be delayed till over nine months after planting. Harvesting may commence between seven and eight months after planting. The tuber length varies between 30 and 40 cm, but may be as long as 72 cm. The harvested tubers are stored by tying them to posts, in barns, under trees, or on platforms. Staggered harvesting is also sometimes practised as a way of increasing the storage period through extension of harvest duration.

Maintenance

The crop is maintained by yearly planting or by retaining them on the field.

Special attributes and uses

The tubers are very palatable and are important for food security. Some cultivars are also used for certain festivals. They are fried, boiled as *ampesi*, or pounded into *fufu*.

Bitter yam (D. dumetorum)

Ecology

The crop is grown in home garden agroforestry, orchards, agroforest non-home gardens, and annual cropping land-use systems.

Planting time and land preparation

These are the same as for the bush yam.

Planting materials

The main sources of planting materials are heirloom cultivars and materials purchased from the market or obtained from other farmers. Like other yam species, the head is the most important planting material, followed by cuttings and small tuber setts. The size of planting materials varies from 200 to 500 g with 250 g as the modal size.

Land preparation and planting

Bitter yams are grown on mounds constructed over planting holes measuring 30 cm to over 50 cm in diameter and 20 cm to 30 cm deep. They are planted as the first crop in a mixed crop system, but no special order is required. The setts are planted 10 cm deep. After planting the mound is capped with dried leaves. The yam setts may be planted directly or nursed and transplanted following the standard nursery procedures.

Staking

The crop is normally staked using live stakes supported by leaders. Staking is done when the vine is between 60 and 100 cm long. But this may be done earlier or later. The distance from a live tree stake to the mound varies between 100 and 230 cm.

Harvesting and storage

The tubers are ready for harvest five to six months after planting and nine months after planting. Harvested tubers are usually short, measuring between 10 and 30 cm in length. Tubers are stored by delaying harvesting. Once harvested, they remain palatable for only one day.

Maintenance

The crop is maintained by yearly planting or by retention in the field.

Special attributes and uses

The tubers are grown for sale or for food security. Some bitter yam varieties are said to be very palatable. The bitter yam is solely used for *ampesi*.

Aerial or bulbil-bearing yam (D. bulbifera)

Ecology

Aerial yams are mostly found in annual cropping systems. They are planted between March and May.

Land preparation and planting

Both freshly cleared and previously cropped lands are used. The crop is grown in a mound or in a hole on the flat. The planting holes are 10–50 cm deep and 20–30 cm wide. Small whole tubers weighing 200–250 g or aerial bulbils are used as planting materials. The sett is planted 10–15 cm deep and mounded over with or without capping. The planting materials are usually nursed following the techniques outlined for the other yams. Duration of nursing varies between three and 12 weeks.

Staking

The crop is staked using the standard staking techniques described above.

Harvesting and storage

Aerial yams mature in nine months after planting. The bulbils are small, measuring 1–9 cm in length with very smooth skin. Harvested bulbils are stored in barns or stored covered under trees.

Maintenance and utilization

The aerial yam is maintained by yearly planting. The bulbils are boiled for *ampesi*. The underground tuber is usually used only as a planting material.

Traditional methods of conservation of diversity in yams in southern Ghana

With the exception of some white yam cultivars, notably *bayere fitaa*, *puna*, *labrokor*, and *dundu banza*, the bulk of the yams cultivated in the zone are heirloom varieties handed down for decades. Over the years the farmers have developed several strategies for ensuring the maintenance of these land varieties in spite of uncertainties in the weather. Table 7.1 summarizes some of the strategies employed.

Table 7.1 Yam maintenance strategies

Strategy	Frequency of practice (%)
Regular replanting of heirloom varieties	100
High-priced varieties and varieties planted in home gardens under intensive care	100
Varieties at risk of disappearing are actively sought out by farmers on germplasm collection trips	37.3
Self-regenerated plants from naturally dispersed seeds produced by flowering varieties are protected	80
Tubers are milked and the vines replanted to produce planting material	100
Careful storage of planting materials on raised platform barns, etc. to ensure viability	70
Exchange of planting materials between farmers during farmers' gatherings, etc.	
Lodging of germplasm with relatives living outside immediate vicinity as insurance against lost of cultivars under unfavourable conditions	37.3
Reintroduction of some cultivars/land-races from original sources	37.3
Management of some heirloom varieties as perennials in home gardens and the main farms	100

Summary and conclusion

The rich diversity of yams held by farmers in southern Ghana owes its existence to the intricately woven socio-cultural practices as well as management systems adopted by the farmers. Specific varieties are required for special cultural activities, thereby necessitating the continued cultivation of such varieties. Also many customs are woven around yam harvest which control the timing of the sale and consumption of yam tubers after harvest that may indirectly contribute to maintenance of the yams. Another contributing factor to the maintenance of yam agrodiversity is that specific yam types are required for various dishes consumed by the farm family. This demands that the different types are continually planted for family needs.

Farmers exploit the different ecological niches on their holdings for maintenance of a wide diversity of yams. The different yam varieties are grown under different land-use systems. Yam species such as bush yams, aerial yams, and trifoliate yams that are adapted to forest conditions are grown under agroforestry systems or sometimes in conserved forests. On the other hand, water yams, Chinese yams, and yellow yams are preferentially grown under more sunny conditions in land-use systems such as annual cropping systems or in non-agroforestry home gardens. The different land-use systems are also exploited for the maintenance and conservation of the yams. Hence under agroforestry and forest fallows, seeds of some bush yams are regenerated from volunteer seeds and new volunteer plants are also produced from bulbils of bulbil-bearing yams or from tuber pieces of yams that are inadvertently left in the ground.

Many species- or type-specific cultural practices have been developed to optimize the yields of the different yam types. For example, in the case of mounding, both yam type and soil structure are considered. On deep, friable soils where moisture is not limiting, bush yams are planted in planting holes without mounding, or with small mounds. On less favourable soils larger mounds are used. Yellow yams are typically grown in very large mounds.

Farmers have of their own volition developed various strategies for maintaining the diversity of yams cultivated. Sometimes methods of farm management such as the use of group labour (the *Nnoboa* system) expose farmers to the different varieties grown by different farmers and permit the exchange of germplasm. Other practices, such as exchange of planting material for farm labour, also help to disseminate and thereby maintain yam diversity.

Despite the various strategies adopted by farmers for the maintenance of yam diversity, erosion of diversity is still a major problem. Many factors contribute to this. For instance, planting material distribution is not well organized. The bulk of it is constituted of heirloom or other common varieties that are under total control of the farmers who replant and maintain them. As commonly occurs in many asexually propagated plants when improperly handled, viruses and other disease organisms accumulate in them, leading to a decline in performance with concomitant loss of planting material.

There is the need to assist farmers in the maintenance of clean planting material. Research organizations may participate in the cleaning and maintenance of buffer stocks of the heirloom varieties for replenishing farmers' stocks when necessary.

Also, the rich experience gained by the farmers in the cultivation of the various yam types may be exploited for the maintenance of yam agrodiversity by forging partnerships between farmers and scientific institutions engaged in germplasm conservation to enhance *in situ* conservation. Such a practice is already in existence between the PLEC farmers of southern Ghana and the Plant Genetic Resources Centre of Ghana.

REFERENCES

Abbiw, D. K., *Useful Plants of Ghana: West African Uses of Wild and Cultivated Plants*, London: Intermediate Technology Publications, Kew: Royal Botanical Gardens, 1990.

Blay, E. T., "Diversity of yams in PLEC demonstration sites in Southern Ghana", *PLEC News and Views*, No. 20, 2002, pp. 25–35.

Degras, L., *The Yam. A Tropical Root Crop*, London and Basingstoke: Macmillan Press 1993.

Kranjac-Berisavljevic, G. and B. Z. Gandaa, "Sustaining diversity of yams in northern Ghana", *PLEC News and Views*, No. 20, 2002, pp. 36–43.

Irvine, F. R., *West African Crops*, London: Oxford University Press, 1979.

8

Sustaining diversity of yams in northern Ghana

Gordana Kranjac-Berisavljevic and Bizoola Z. Gandaa

Introduction

In northern Ghana, yam cultivation is widespread in almost all the settlements. A survey carried out by the northern Ghana PLEC team in 2001 indicates that about 75 per cent of farmers in Bongnayili-Dugu-Song, the main PLEC demonstration site in northern Ghana cultivate yam (Map B). On an average, about five yam types are found on every yam farmer's field.

Blench described yam species cultivated in West Africa as "one of the crops at the dynamic frontier between wild and domestic" (Blench, 1997:1), since there is much information about yam which has not yet been investigated by researchers. He further argues that such crops have much to offer in terms of food security, since people continue to cultivate them. It is generally understood by farmers that new yam species are brought from the bush or forest (*YoÂo* in the Dagbani language).

Yam domestication practices in northern Ghana

Newly discovered yam seedlings in the forest are sometimes "bitter" in taste or tasteless. Fetish priests in northern Ghana commonly domesticate new yam types brought into the community in shrines, even though individual farmers can cultivate yam brought from the forest on their own. New yam types are cultivated about four to seven years before being given a name, which is often descriptive in nature. Some of the shrines are located at Siiyare, Birikum, and Gambugu, all settlements in the Nanumba district of the northern region of Ghana. The essence of cultivation in the shrine is to receive blessing of the newly discovered yam types.

Cultivation methods

Several yam species are grown in Bongnayili-Dugu-Song, with details presented in Table 8.1.

Yam is normally grown as the first crop in a rotation after the land has been cleared. Usually little or no fertilizer is used in yam cultivation, even though the crop responds favourably to application of phosphorus and potassium. Yam performs best in deep, well-drained loamy soils and is frequently grown in a mixture with other crops such as millet, sorghum, maize, rice, cowpea, and other crops of savanna and transitional ecological zones. However, a clear relationship exists between cultivation of yam as a sole crop and high yield (Kowal and Kassam, 1978).

In northern Ghana, land preparation for yam cultivation involves use of the cutlass and hoe for clearing bush followed by burning of the residue. The

Table 8.1 Yam accessions in a PLEC demonstration site in northern Ghana.

Accession no.	Local name	Yam species
1	*Digi*	*D. rotundata*
2	*Bombe-tingye*	*D. rotundata*
3	*Kpuringa*	*D. rotundata*
4	*Laabako*	*D. rotundata*
5	*Chamba*	*D. rotundata*
6	*Manchisi*	*D. alata*
7	*Kpuno*	*D. rotundata*
8	*Zuglanbo*	*D. rotundata*
9	*Kan-gbaringa*	*D. rotundata*
10	*Chenchito*	*D. rotundata*
11	*Fugla/fugra*	*D. rotundata*
12	*Zong*	*D. rotundata*
13	*Baamuyegu*	*D. rotundata*
14	*Liilya*	*D. rotundata*
15	*Kiki*	*D. rotundata*
16	*Dakpani*	*D. rotundata*
17	*Baayeri*	*D. rotundata*
18	*Goenyeni*	*D. rotundata*
19	*Mogni-nyugo*	*D. rotundata*
20	*Gun-gonsalli*	*D. rotundata*
21	*Nyuwogu*	*D. alata*
22	(*Nkanfu*?)	*D. bulbifera*

Source: PLEC field survey, 1999–2001

essence of this practice, according to farmers, is to destroy termites and other insects that attack yam tubers. Bullock-drawn ploughs are used to break the land after the first rains. Raising of mounds, 0.4–0.6 m high, spaced 1.2–1.5 m apart, follows this operation. According to farmers, soil suitability for yam cultivation is evident in the presence of wild yam types in the area, which is an indication of deep soils. Planting is done two days after the mounds are raised and the planted seed-yam is mulched to reduce the soil water loss and protect the seed-yam from solar radiation.

Two weeks after sprouting, the yam vines are either staked or not staked depending on the yam type (Kranjac-Berisavljevic and Gandaa, 2000). Pricking is done 10 weeks after sprouting. During a typical annual cultivation cycle, yam is usually the last crop, before cassava, to be harvested.

Generally, harvested yam is stored in huts constructed from grass in the yam farms. This allows ventilation and also minimizes losses through transportation. Another common practice among farmers is to carry the yams to their homes where they are stored in a hut constructed with mud and roofed with grass. In these huts, care is taken not to place the yam tubers on top of each other.

Collection of yam types at Bongnayili-Dugu-Song PLEC demonstration site

At Bongnayili-Dugu-Song, the main PLEC demonstration site in northern Ghana, practising researchers and farmers work hand in hand on yam conservation and multiplication, respecting each other's methods of gathering information while reaching conclusions in a collaborative manner.

Twenty-two yam accessions from northern Ghana were analysed during 2001 with the help of PLEC researchers at the Crop Science Department of the University of Ghana, Legon, using esterase and total protein banding patterns in order to determine which of the yam types are similar to each other. The results of these analyses are shown as Table 8.1 and Figures 8.1 and 8.2. Accessions numbered 1 and 4 in Figure 8.1 (*digi* and *laabako*) appear to be quite similar according to esterase banding patterns, while accessions 5, 6, 7, 9, and 10 (*chamba*, *manchisi*, *kpuno*, *kan-gbaringa*, and *chenchito*, respectively) appear to be virtually the same. The same is true for accessions numbered 12, 16, and 18 (*zong*, *dakpani*, and *goenyeni*), accessions 3 and 13 (*kpuringa* and *baamuyegu*), 14, 17, and 15 (*liilya*, *baayeri*, and *kiki*) and 19, 20, and 21 (*mogni-nyugo*, *gun-gonsalli*, and *nyuwogu*).

The dendogram presented in Figure 8.2 shows the links among yam types based on total protein content. Yam type accessions numbered 13 and 19 (*baamuyegu* and *mogni-nyugo*) and 20 and 21 (*gun-gonsalli* and *nyuwogu*) also appear the same, while the accessions numbered 1 and 2 (*digi* and *bombe-tingye*), 7 and 8

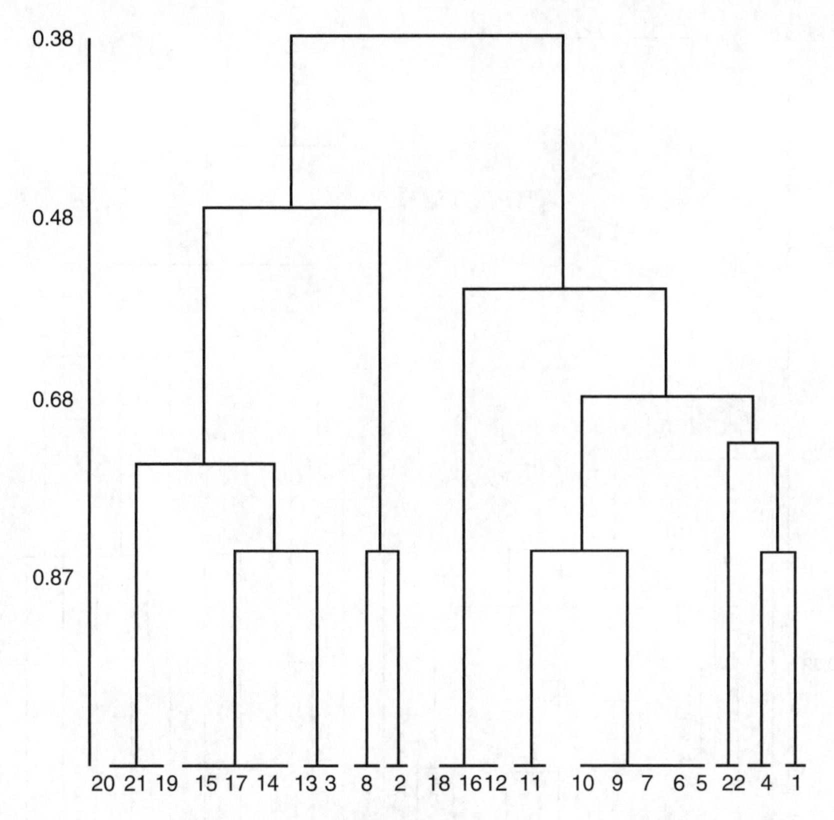

Figure 8.1 Relationships between northern yam types cultivated in a PLEC demonstration site in northern Ghana: Dendogram of 22 accessions of yams based on esterase binding pattern

(*kpuno* and *Zuglanbo*), 9 and 10 (*kan-gbaringa* and *chenchito*), 12 and 14 (*zong* and *liilya*), and 16 and 17 (*dakpani* and *baayeri*), appear to be quite closely related on the basis of total protein.

Field collection and characterization of yam types

Farmers working on the demonstration site further identified yam types on the field using the following criteria:

- nature of vines (staked or cripping)
- leaf characteristics
- tuber characteristics.

Farmers also differentiated wild yam from those domesticated by the direction (clockwise or anti-clockwise) of circummutation of the vines.

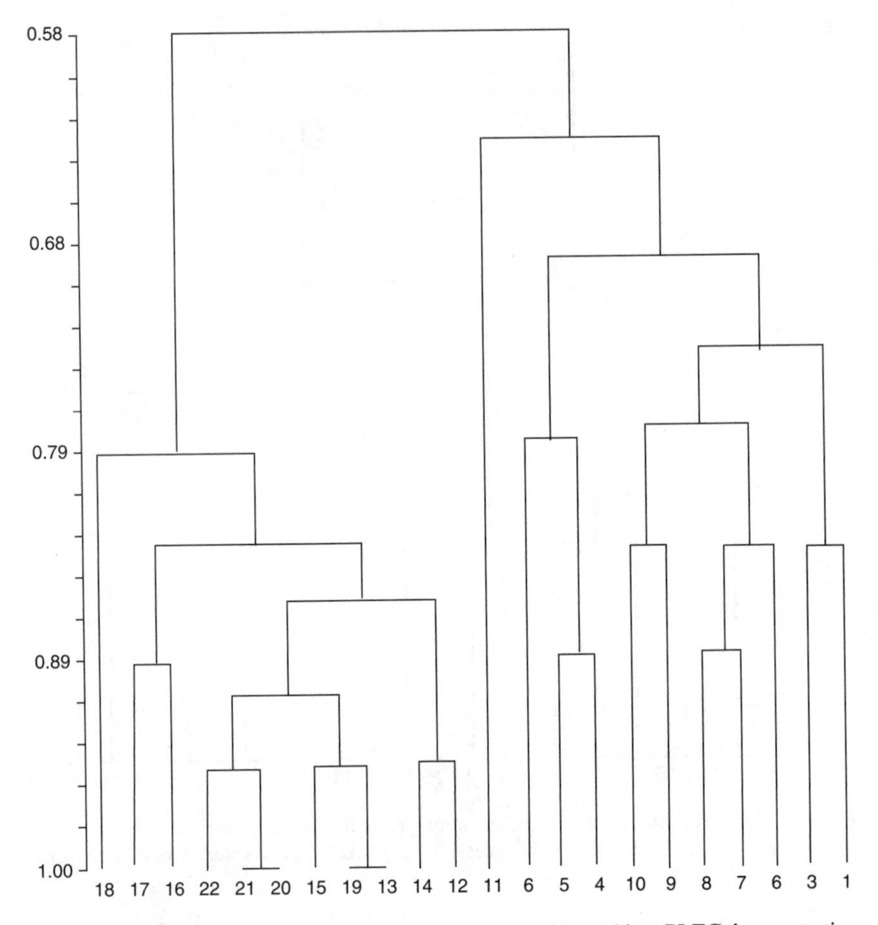

Figure 8.2 Relationships between northern yam types cultivated in a PLEC demonstration site in northern Ghana: Dendogram of 22 accessions of yams based on total protein

In line with PLEC methodology, a yam type collection farm has been created and managed by PLEC expert yam farmers at Dugu. The aims of creating the farm were to:

- collect new yam types identified by farmers
- multiply yam types and sell at an affordable price to farmers
- serve as a germplasm bank.

Detailed morphological characterization of the 22 identified yam types presented in Table 8.1 was carried out during the year 2000. The results of these observations were confirmed in the 2001 cultivation season. Characteristics of each yam type were observed at regular intervals. Results are presented in Tables 8.2 and 8.3. About 11 main characteristics were observed throughout the season for all the yam types. Sprouting percentage was found to be very

Table 6.2 Yam types observed during the 1999 growing season at Bongnayili-Dugu-Song

Yam type	Origin (according to farmers)	Main use	Agronomic characteristics	Other remarks	Length of tuber (cm)	Weight of tuber (kg)
1. *Digi*	–	NA	NA	NA	NA	NA
2. *Bombe-tingye*	Dagbon	Consumed locally, sold as cash crop	Stacked, not pricked	–	45	1.65
3. *Kpuringa*	Dagbon	Consumed locally (sliced and roasted)	Stacked plant	Eaten mostly by children (small-size tubers)	21	0.8
4. *Laabako*	Dagbon	Consumed locally (as *fufu*, sliced yam), also sold as cash crop	Normally grown stacked	Women prefer this type, since it is easier to cook. It is also used to perform yam sacrifices, due to its early-maturing qualities	28	1.8
5. *Chamba*	Dagbon	Used for *fufu*	Late stacked	–	36	0.8
6. *Manchisi*	Dagbon	Consumed locally as sliced or roasted yam, also sold as cash crop	Late stacked	–	15	0.7
7. *Kpuno*	Dagbon	Consumed locally (as *fufu*, sliced yam), also sold as cash crop	Normally grown stacked	–	27	1.7
8. *Zuglanbo*	NA	NA	NA	NA	NA	NA
9. *Kan-gbaringa*	NA	Consumed locally as *fufu*, or sliced yam	Stacked and prickable	–	25	1.3
10. *Chenchito*	NA	Consumed locally (as *fufu*), also sold as cash crop	Stacked late in the season	This type of yam is eaten at funerals and festivals	30	1.9
11. *Fugla/fugra*	NA	Consumed locally (sliced and roasted)	Cripping plant	Rarely sold for cash	29	1.9
12. *Zong*	Dagbon	Consumed locally (sliced and roasted)	Cripping plant	–	NA	NA

Table 8.2 (cont.)

Yam type	Origin (according to farmers)	Main use	Agronomic characteristics	Other remarks	Length of tuber (cm)	Weight of tuber (kg)
13. *Baamuyegu*	Dagbon	Consumed locally	Cripping plant	–	20	0.1
14. *Liilya*	Yendi area	Cash crop, sometimes eaten at homes	Stacked plant	–	19	0.6
15. *Kiki*	Yendi area	Cash crop, sometimes eaten at homes	Stacked plant	–	NA	NA
16. *Dakpani*	Yendi area	Cash crop, sometimes eaten at homes	Stacked plant	–	15	0.3
17. *Baayeri*	Dagbon	Consumed locally as sliced yam, and *fufu*, Sold as cash crop	Stacked, also pricked	–	19	0.2
18. *Goenyeni*	Dagbon	Consumed locally (sliced and roasted)	Cripping plant	–	28	3.1
19. *Mogni-nyugo*	Yendi area	Consumed locally (sliced and roasted, as *fufu*), also used as cash crop	Cripping plant	–	NA	NA
20. *Gun-gonsalli*	Dagbon	Consumed locally as sliced yam, good for *fufu* or roasted. Not sold	Cripping plant	–	14	0.9
21. *Nyuwogu*	Dagbon	Consumed locally (sliced and roasted)	Stacked plant	–	25	2.7
22. *Nkanfu*	South Ghana	Consumed locally as sliced yam	Stacked	Due to bitter taste, it is cooked and left overnight to be eaten the following morning	6	0.1
23. *Friginli*	Dagbon	Consumed cooked and roasted, also cash crop	Stacked	Aerial yam	10	0.1
24. *Kukulga*	Dagbon	Consumed cooked and roasted, also cash crop	Stacked	Aerial yam	37	1.4

Table 8.3 Vegetative characteristics of yam types collected at Bongnayili-Dugu-Song

No.	Yam local name (Dagbani)	% sprouting	Length of petiole (cm)	Distance between the nodes (cm)	Leaf length (cm)	Leaf width at mid-region (cm)	Number of stems/tuber	Leaf area index	Circummutation	Tuber weight (kg)	Tuber length (cm)
1.	*Digi*	92	2.9	12.3	7.2	5.4	4	78.8	Clockwise	0.35	20.0
2.	*Bombe-tingye*	70	4.5	22.5	8.3	6.0	1	274.5	Clockwise	0.55	25.0
3.	*Kpuringa*	70	5.0	10.0	7.0	5.0	6	239.4	Clockwise	0.29	14.0
4.	*Laabako*	89	6.5	16.5	10.0	8.0	4	526.5	Clockwise	1.91	26.0
5.	*Chamba*	80	7.0	12.0	15.0	9.0	1	106.3	Clockwise	1.00	11.0
6.	*Manchisi*	64	10.7	12.5	10.5	7.3	3	324.8	Clockwise	0.61	9.0
7.	*Kpuno*	86	9.0	16.5	14.0	13.0	2	100.8	Clockwise	2.30	31.0
8.	*Zuglanbo*	NA	NA	NA	NA	NA	NA	NA	NA	NA	NA
9.	*Kan-gbaringa*	40	6.0	13.3	10.3	8.3	6	122.3	NA	0.75	22.0
10.	*Chenchito*	82	9.5	17.0	10.5	7.5	8	1,251.7	NA	0.90	27.0
11.	*Fugla*	70	4.0	8.0	9.0	4.0	4	126.9	NA	0.56	17.0
12.	*Zong*	60	6.0	17.0	10.0	6.0	1	125.1	NA	0.25	13.0
13.	*Baamuyegu*	80	6.0	12.0	7.5	5.6	2	110.4	NA	0.15	8.5
14.	*Liilya*	52	5.7	18.3	12.6	8.2	2	96.3	Clockwise	0.85	15.5
15.	*Kiki*	80	5.0	10.0	8.0	6.0	3	336.0	Clockwise	0.60	18.5
16.	*Dakpani*	56	5.0	16.0	11.0	6.0	4	92.5	Clockwise	0.30	11.0
17.	*Baayeri*	70	3.5	11.5	8.5	5.5	2	188.4	Clockwise	0.92	25.5
18.	*Goenyeni*	84	9.0	11.3	12.0	8.0	3	211.9	Clockwise	0.50	18.0
19.	*Mogni-nyugo*	42	8.0	20.3	10.0	6.5	2	106.3	Clockwise	0.90	19.0
20.	*Gun-gonsalli*	60	5.0	23.0	9.0	5.0	2	99.6	Clockwise	0.71	18.0
21.	*Nyuwogu*	84	7.0	9.0	13.0	10.0	8	551.6	Clockwise	0.06	9.0
22.	*Nkanfu*	NA	1.2	5.1	3.2	3.0	1	NA	NA	NA	NA
23.	*Friginli*	86	10.0	17.0	15.0	15.3	1	124.6	NA	0.20	9.5

Source: PLEC field survey, 2001

variable from one yam type to another, ranging from about 40 per cent up to about 92 per cent (*digi*). This variability may be explained partly by the fact that various yam types require a different planting time, as well as planting depth. These issues require further investigation.

Length of petiole was also very variable, ranging from 1.2 to 10.7 cm. Similarly, other morphological characteristics show great variability from type to type (Table 8.3). Almost all the yam types also showed clockwise circummutating. Regretfully this type of information was not available for the *nkanfu* yam type.

Great dissimilarity in tuber weight and length at the end of the growing season corresponded well with already observed variations in vegetative characteristics during the season. The tuber weight ranged from 0.06 (*nyuwogu*) to 2.30 kg (*kpuno*). The number of stems also ranged from one (*zong*, *friginli*, *chamba*) to about eight (*nyuwogu*, *chenchito*).

Chemical content

In order to complement effectively the traditional knowledge and field observations, laboratory analysis of the chemical content of 22 yam types was also conducted with the assistance of the Food Research Institute Laboratory, Accra, in 2001.

Data on seven parameters obtained during the tuber analysis and a summary of the results are presented in Table 8.4.

Moisture content was variable for various yam types. *Manchisi* had the highest moisture content (71.9 per cent), and *chamba* (50.2 per cent) had the lowest moisture content.

Coursey (1976) reported that ash content of yams is high, but varies from species to species. Apart from *zong* (1.6 per cent), *nyuwogu* (1.6 per cent), and *friginli* (1.5 per cent), which had the highest ash content, the rest of the yam types from Bongnayili-Dugu-Song had ash contents ranging from 0.9 to 1.4 per cent.

Contrary to some earlier reports indicating relatively high protein contents in yam, among the 22 types collected, the protein content ranges from 1.6 (*manchisi*) to 4.1 per cent (*kan-gbaringa*), which is rather low. Starch content was very variable, ranging from 6.0 (*mogni-nyugo*) to 27 per cent (*kpuringa*).

Tindal (1983) reports that 100 g yam has about 1 mg phosphorus and 0.8 mg of iron. Yams at Bongnayili-Dugu-Song have a phosphorus level content ranging from 29.2 mg/100 g (*gun-gonsalli*) to 3.8 mg/100 g (*kukulga*), iron content ranging from 11.1 mg/100 g (*zong*) to 0.5 mg/100 g (*gun-gonsalli*, *friginli*, *baamyegu*, *digi*, *kpuringa*, and *liilya*), and a variable calcium content

Table 8.4 Percentage moisture, ash, protein, starch, and minerals in 22 yam types grown in Bongnayili-Dugu-Song

Yam type	% moisture	% ash	% protein	% starch	mg/100 g iron	mg/100 g calcium	mg/100 g phosphorus
1. *Digi*	63.4	1.1	3.2	7.9	0.5	19.5	7.6
2. *Bombe-tingye*	51.3	1.1	2.9	12.2	0.8	14.3	3.7
3. *Kpuringa*	56.1	1.2	1.8	27.0	0.5	25.8	6.6
4. *Laabako*	52.6	1.0	1.7	6.6	0.8	24.3	4.2
5. *Chamba*	50.2	1.0	3.2	26.6	1.1	20.7	26.4
6. *Manchisi*	71.9	1.7	1.6	13.6	1.1	16.6	20.4
7. *Kpuno*	51.6	1.4	3.3	11.7	6.0	20.2	6.8
8. *Zuglanbo*	NA	NA	NA	NA	NA	NA	NA
9. *Kan-gbaringa*	57.9	1.2	4.1	14.3	1.1	11.1	12.3
10. *Chenchito*	60.3	1.0	2.6	23.0	0.7	8.5	5.3
11. *Fugla/fugra*	58.4	0.9	2.3	13.6	1.1	16.6	18.0
12. *Zong*	61.0	1.6	2.2	8.0	11.1	22.2	8.7
13. *Baamuyegu*	65.0	0.9	2.2	18.6	0.5	9.5	9.8
14. *Liilya*	61.2	1.2	3.5	19.8	0.5	9.5	4.9
15. *Kiki*	60.0	1.3	1.7	9.4	1.0	16.4	9.3
16. *Dakpani*	57.0	1.2	2.3	11.4	0.8	17.0	6.0
17. *Baayeri*	65.2	1.3	2.4	18.3	1.3	6.3	9.9
18. *Goenyeni*	59.6	1.6	4.0	8.9	1.0	12.5	23.6
19. *Mogni-nyogo*	72.0	1.0	2.4	6.0	0.9	16.4	2.3
20. *Gun-gonsalli*	60.2	1.3	2.7	15.7	0.5	30.5	29.2
21. *Nyuwogu*	67.1	1.6	2.9	10.9	1.0	21.0	31.1
22. *Friginli*	70.8	1.5	2.6	10.5	0.5	7.1	9.0
23. *Kukulga*	55.7	0.9	4.0	11.1	1.2	10.6	10.8

Source: Analysis of data from PLEC field survey, 2001

ranging from 30.5 mg/100 g (*gun-gonsalli*) to 6.3 mg/100 g (*baayeri*). All these values are far in excess of the values reported by Tindal (1983).

Erosion of yam types

Table 8.5 outlines the factors that, reportedly, lead to erosion of the yam types. The erosion of yam types might be accelerated by social, cultural, and economic factors. For example, farmers believe that the *baamuyegu* yam type has spiritual powers and is thus the leader of all yam types. It is believed that it leads the other yam types spiritually during the night (to drink water). When this type of yam increases on a farm, there will be a struggle for "leadership" among the yams, and the farmers will lose all other yam types. Thus, only experienced farmers keep the *baamuyegu* type of yam on their farms.

Table 8.6 indicates the yam types maintained by farmers and the reasons why they are kept. The most common yam types kept are *goenyeni*, *kpuringa*, *laabako*, *kpuno*, and *baayeri*. Prosperous farmers are those with large farms of *Laabako* and *Kpuno* yam types. There is a competition among the farmers to cultivate these yam types.

Table 8.5 Factors leading to erosion of some yam types in a PLEC demonstration site in northern Ghana

Factor	Type of yam
Declining soil fertility	*Liilya, chamba, zong, kukulga, bombe-tingye*
Spiritual	*Baamuyegu*
Drought	*Nkanfu, kan-gbaringa*
Difficult to store	*Manchisi, gun-goansalli, mogni-nyuga*
Labour intensive	*Digi, dakpani*

Source: PLEC field survey, 2001

Table 8.6 Factors leading to maintenance of some yam types in a PLEC demonstration site in northern Ghana

Factors	Yam type
Cash source	*Laabako, kpuno, friginli, baayeri*
Festivals and funerals	*Chenchito, kpuringa, friginli*
Hunger crop	*Kpuringa, goenyeni, zuglanbo, fugla*
Sacrifice (yam festival)	*Laabako*
Easy preparation	*Laabako, kpuno*

Source: PLEC field survey, 2001

Conclusions

A survey carried out by the northern Ghana PLEC team in 2001 indicates that about 75 per cent of farmers in Bongnayili-Dugu-Song, the main PLEC demonstration site in northern Ghana, cultivate yam. On an average, about five yam types are found on every yam farmer's field. Farmers are able to supply information on the traditional methods of yam domestication, characteristics of the crop such as colour, leaf, stem, and tuber, and also on the usefulness of particular yam types in terms of storability and taste. Farmers have many reasons for growing certain yam types. The reasons range from spiritual to commercial. It is also observed that some yam types growing in the wild are indicators of soil fertility.

The farmers provided the PLEC research team with a yam plantation site where all the collected 22 northern yam types were planted for further observations.

During the 1999–2001 rainy seasons, 22 yam types were identified and collected by farmers. Detailed information on the morphological characteristics and nutritional value of the yam types is now available.

Chemical analysis carried out in 2001 show that the yam types in northern Ghana have a very low protein content. The protein content ranges from 1.6 per cent (*manchisi*) to 4.1 per cent (*kan-gbaringa*). Similarities and differences among the 22 yam types, based on the esterase and total protein banding patterns, are also available.

Phosphorus level content for the same types ranged from 29.2 mg/100 g (*gun-gonsalli*) to 3.8 mg/100 g (*kukulga*), and iron content ranged from 11.1 mg/100 g (*zong*) to 0.5 mg/100 g (*gun-gonsalli, friginli, baamyegu, digi, kpuringa,* and *liilya*). Calcium content ranged between 30.5 mg/100 g (*gun-gonsalli*) and 6.3 mg/100 g (*baayeri*). These values are far in excess of the values reported in the literature by other authors.

Variability and disparity in all the results indicate that much more work needs to be done in order to obtain a clearer picture on nutritional values, genetic relationships, and other characteristics of various traditional yam types cultivated in northern Ghana.

Further collaborative studies are planned to compare the varieties cultivated in northern Ghana with those grown in the other parts of the country, notably the Sekesua-Osonson PLEC demonstration site in southern Ghana (see Chapter 7).

REFERENCES

Blench, R., "Neglected species, livelihoods and biodiversity in difficult areas: How should the public sector respond?", *ODI, National Resources Perspectives*, Vol. 23, 1997, pp. 1–4.

Coursey, D. G. "The origin and domestication of yams in Africa", in J. R. Harlan, J. M. J. de Wet, and A. B. I. Stemlar, eds, *Origin of African Plants Domestication*, The Hague: Morton, 1976, pp. 383–403.

Kowal, J. M. and A. H. Kassam, *Agricultural Ecology of Savanna: A Study of West Africa*, Oxford: Clarendon Press, 1978.

Kranjac-Berisavljevic, G. and B. Z. Gandaa, "Collection of yam types at Bongnayili-Dugu-Song, main PLEC demonstration site in northern Ghana", *PLEC News and Views*, No. 15, 2000, pp. 27–30.

Tindal, H. D., *Vegetables in the Tropics*, London, Macmillan Press, 1983.

9

Conservation of indigenous rice varieties by women of Gore in the northern savanna zone, Ghana

Paul B. Tanzubil, Joseph S. Dittoh, and Gordana Kranjac-Berisavljevic

Introduction

Plant genetic diversity is a key ingredient for sustainable development. Genetic diversity buffers peasant farmers' cultivation practices against environmental hazards (e.g. drought, floods) and changing market conditions as well as pest and disease outbreaks.

Tropical and subtropical Africa is the centre of diversity for a range of crops including the African rice, *Oryza glaberrima*. The grain of *O. glaberrima*-type rice has a high gluten content (Longley and Sellu-Jusu, 1999). It is pleasantly filling when eaten and makes the consumer feel satisfied for a longer period after the meal than *O. sativa*. This type of rice can therefore still be found with many farmers in northern Ghana (Map B), with a great deal of varietal diversity.

African subsistence farmers have traditionally relied on such diversity to ensure the stability of their food production systems. Wild resources have thus continued to be important for food and the livelihood security of the rural poor, including women and children, especially in times of stress such as drought and changing land and water availability, or even ecological change. Unfortunately, when modern varieties replace traditional plant materials, genetic resources are rapidly destroyed.

Since the inception of the green revolution, traditional systems of agriculture in most parts of the world have become threatened. These systems, which are characterized by a high diversity of crops, are largely being replaced by systems that depend on a few commercial crops and sometimes uniform varieties with a very narrow genetic base.

Small-scale farmers, like those in northern Ghana, are often the worst affected by these changes, as they are barely able to purchase the external inputs required under modern production systems. The rural poor generally have less access to land, labour, and capital and thus rely more on the diversity available in their local ecosystems. They have developed diverse land-races and cropping patterns adapted to local climate, social, and cultural situations (M'Mukindia, 1994). Such farmers therefore tend to exhibit some resistance to changes in production practices, and remain rather impervious to extension messages that stress high external input agriculture.

A conspicuous example of this type of traditional agriculture has been the reaction of farmers to improved crop varieties introduced to them from outside their communities. The results of such introductions have seldom been spectacular. Many farmers continue to grow their own crop varieties, which they are familiar with.

The women farmers of Gore, a village in the Bawku area of the Upper East region of Ghana, provide ample testimony to this trend (Map B). They maintain and continue to cultivate some traditional rice varieties, in spite of a proliferation of improved rice types introduced into the community by research and development workers. Farmers' varieties are uniquely adapted, genetically diverse, and possess characteristics evolved over time through cultivation and selection in their local environments.

The northern Ghana PLEC team of scientists was attracted into a partnership with the Gore women group with a view to approaching the conservation, management, and utilization of the indigenous varieties of rice in a collaborative way. This chapter reports the results of that effort.

Methods and materials

The conservation activities are in progress at Gore, one of the three subdemonstration sites where PLEC work is undertaken in the Bawku area. The site falls within the Sudan savanna agro-ecological zone. It is located about five kilometres from the Manga Research Station of the Council for Scientific and Industrial Research (CSIR). Mean annual rainfall at the site is about 900 mm. The area is characterized by depressions suitable for rain-fed rice cultivation.

Rural women and men farmers are key partners in this participatory action research. This approach is unique because research and extension agents had, in the past, seldom targeted rural women for collaboration in this part of the country. In PLEC's preliminary work in 1998 and 1999, a participatory technology development (PTD) approach was used. The critical ingredient of this approach is a multidisciplinary research team. The team consisted at various stages of an engineer, an agricultural economist, an agronomist/soil scientist, an extensionist, and a crop protectionist/IPM expert.

Another important ingredient of the research approach was the use of participatory rural appraisal (PRA) tools such as transect walks, reconnaissance surveys, agro-ecosystem analysis, and various forms of stakeholder interviews. Using a combination of these tools, information was sought about indigenous rice varieties grown in the valley bottoms by members of the Gore community, with emphasis on women farmers. Interviews and discussions with farmers were held in a very relaxed atmosphere and at their convenience in terms of venue and time. Team members also embarked on transect walks with farmers, during which comments and explanations about indigenous rice varieties were sought. Throughout the studies, the team recognized the farmers as equal partners and, in some cases, farmers were in the lead, effectively dictating the direction and pace of the PTD process. This approach aroused enthusiasm, interest, and active participation in the community, who provided information freely and revealed pertinent issues which otherwise could have been overlooked.

During the first two seasons, much effort was put into observing rice production practices and trying to characterize the different varieties. On-farm trials were also conducted in the 1998 and 1999 rainy seasons to study the characteristics of two indigenous rice varieties and compare their performance with that of two improved varieties. The test indigenous rice varieties, namely *asamolgu* and *asakyira*, were selected for initial evaluation by 18 farmers alongside IR-24 and GR-18, both varieties released by the National Agricultural Research System (NARS). The size of each experimental plot was 5 × 5 m and the seed of each variety was dibbled in the soil at a spacing of 20 × 20 cm. Mineral fertilizer was applied at a rate of 60-40-40 kg NPK per hectare. Basal application was 15-15-15, and later top-dressing was applied with sulphate of ammonia at the rice tillering stage. Yield and other vital data were collected, analysed, and documented for both traditional and improved rice varieties. During the 2001 cropping season the research team tried to characterize the indigenous rice varieties by assisting each woman to grow one or two indigenous rice varieties of their choice. The exercise was organized along the lines of the farmer field school (FFS) approach where groups of farmers are brought together to exchange ideas and learn techniques on the cultivation of a crop. In the case of Gore women, meetings were organized in the rice fields at different stages of development of the rice crop to afford them the opportunity of discussing the rice ecosystem and its challenges. Farmers also made observations on differences in performance of their respective fields and tried to assign reasons for such differences in a participatory manner. The researchers also collected agronomic data on each plot, especially maturity period, height, number of panicles, and paddy yield of each variety.

From the year 2001, steps were taken to replicate the experience with the Gore women at Kusanaba (a village about 70 km from Gore), where women were already involved in some form of *in-situ* conservation of indigenous vegetable and rice varieties.

After reconnaissance surveys were conducted by the whole northern Ghana team of PLEC scientists, different subteams carried out specific tasks at the site including mapping, plot demarcation, topical interviews, and organization of participatory learning exercises. Information on the types of indigenous rice varieties, their characteristics, and the agronomic practices adopted by farmers growing them were collected. The women, to multiply seed of six indigenous rice varieties most preferred by them, established a community seed plot. It is hoped that this would lead to wider cultivation of indigenous rice varieties and hence contribute to the conservation of agrobiodiversity in the communities at the PLEC sites. The scope for extending these experiences to new communities is being assessed.

Results and discussion

The participatory approach to research adopted by the team revealed the central role of women farmers in conserving the indigenous rice varieties. While most men farmers at Gore had forgotten the names of the indigenous rice varieties, the women could name up to 12 indigenous varieties. Many of the women actually continue to cultivate the local varieties, while most men have replaced them with improved ones. More women were involved in rice cultivation than men, but their rice fields tended to be smaller (Table 9.1).

Women rice farmers also tended to have their farms situated in disadvantaged portions of the toposequence compared to those of men. Their fields, however, are usually better managed than those of men, probably due to the smaller sizes and fewer crops they generally grow.

Gore farmers have names for all the indigenous rice varieties they cultivate. The names describe either the origin of the varieties or their peculiar characteristics. For example, some indigenous rice varieties have been named after farmers who first introduced them into the community. Among them are such varieties as "Peter" and "Mr Moore". The names of other varieties give indications of their

Table 9.1 Differences in production practices between men and women farmers at Gore (2000 field survey)

Item	Women farmers	Men farmers
Location	Mainly upland	Largely lowland
Field size	0.1–0.3 ha	0.3–2 ha
External input use	2–3 bags/ha NPK	Up to 5 bags/ha NPK
Field management	2–4 weedings	1–2 weedings
Varieties grown	Largely local	IR-24, GR-19
Use of produce	Food and sale	Mainly sold
Yield of grain/ha	10–30 bags	12–38 bags

origin. *Agona* is the name of a variety originating from a town with the same name in the Ashanti region of Ghana. Sometimes the size and shape of the grain give a name to the variety. For instance, *agongula* means short-grained rice. Other names describe the colour of the paddy or milled grain, as in *musabelig*, which refers to the "black" (dark) colour of the husk. All these names indicate that the women farmers have a lot of information about the indigenous rice varieties that they grow.

Farmers mentioned some desirable attributes of indigenous rice varieties, which make them superior to modern varieties. These include the following:

- short cooking time
- indigenous rice varieties do not spoil when cooked and left overnight as many traditional homes in the Upper East region do
- few ingredients are needed for preparation, as they can be cooked with only salt and pepper, but still are very tasty
- they are better suited for making traditional dishes like *waakye* (cooked rice and beans) and rice balls
- indigenous rice varieties are good for weaning babies (i.e. baby food)
- indigenous rice varieties are less prone to shattering in the field, even when harvesting is delayed
- they perform better and give higher yields than improved varieties under low-input technology, especially without mineral fertilization
- they also give higher yields under adverse conditions such as drought, pest, and disease outbreaks
- straw of indigenous rice varieties is preferred by livestock
- indigenous rice varieties are easier to process under the local conditions, usually by pounding manually with the aid of mortar and pestle.

The farmers at Gore identified 13 indigenous rice varieties grown over the years. The characteristics and distinguishing features of each local variety are summarized in Table 9.2. From Table 9.2, it is clear that farmers' stock of varieties keeps changing with time, thus increasing the gene pool within the community. They may accept new introductions from the outside, but nevertheless strive to maintain the varieties they are familiar with. This tendency explains the continuous cultivation of some varieties for over 30 years within the same locality.

Farmers also gave good reasons for continuous cultivation of some indigenous rice varieties that otherwise have poor agronomic qualities and low market value. For instance, *agongula* and *musabelig* are grown largely for local consumption as they have low market value. The varieties with white grain, on the other hand, command good prices at the urban markets and are cropped largely for sale.

From the farm trials it was found that only one of the improved varieties, IR-24, significantly out-yielded the indigenous rice varieties (Table 9.3). There were no significant yield differences among GR-18, *asamolgu* and *asakira*. *Asamolgu* even out-yielded GR-18. This showed that indigenous rice varieties could give high yields under improved technology such as the use of mineral fertilizer. The yield

Table 9.2 Characteristics of indigenous rice varieties grown by women farmers at Gore in the Bawku area of Ghana as mentioned by farmers

IRV	Ecology	Duration (days)	Paddy colour	Grain colour	History of cultivation (years)	Yield potential (tonne/ha)
Asakira	Upland	90	Brown	White	20	2.5–3.0
Asamolgu	Hydromorphic	90	Brown	White	20	2.5–3.0
Nagamui	Upland	90		Red	20	2.5–3.0
Santie	Upland	90	Straw	White	10	2.5–3.0
Agonsanga	Upland	90	Straw	White	6	3.0–3.5
Peter	Upland	90	Straw	Dirty white	5	2.5–3.0
Abunga	Lowland	115–120	Red	White	25	3.0–3.5
Agona	Lowland	115–120	Brown	White	20	3.0–3.5
Agongula	Lowland	115–120	Straw	Red	10	3.0–3.5
Mr Moore	Lowland	115–120	Straw	White	20	3.0–3.5
Mendi	Lowland	90	Brown	White	5	3.0–4.0
Worigaworiga	Upland	NA	Straw	Spotted red	6	2.5–3.0
Musabelig	Upland	NA	Dirty straw	Red	30	2.5–2.5

Table 9.3 Yield of rice varieties in on-farm trials at Gore (mean for 1998 and 1999 growing seasons)

Variety	Yield of paddy (tonnes/ha)
IR-24	3.20
GR-18	2.50
Asamolgu	2.90
Asakira	2.50
LSD (5%)	0.67
CV%	25.00

Note: Numbers of farmers for the trials in 1998 and 1999 are 15 and 18 respectively.

levels recorded during the trials tallied very well with those given by the women farmers during the initial interactions, confirming that farmers know the characteristics of indigenous rice varieties very well.

From the 2001 observations, the good performance of the popular indigenous rice varieties was confirmed. Yields of the indigenous rice varieties were comparable to that of the improved variety, GR-18, and ranged between one and five tonnes/ha (Table 9.4).

In fact, four indigenous rice varieties (Peter, *asamolgu*, *santie* and *agonsanga*) yielded higher than the improved variety.

Table 9.4 Records on cultivation of indigenous rice varieties by Gore women farmers, 2001

Variety	Number of farmers	Hectarage		Panicles/m^2		Days to maturity	Plant height at harvest (CM)		Grain weight (tonnes/ha)	
		Min.	Max.	Range	Mean		Range	Mean	Range	Mean
Agongbilla	1		0.11	38	39	125	115–117	116		1.33
Agonsanga	1		0.07	24	34	132	93–110	102		4.80
Peter	2	0.03	0.07	21–48	34.3	78	95–136	117	1.07–4.80	2.94
Abonga	2	0.02	0.04	21–35	28.2	116	100–115	111	1.33–1.87	1.60
Awariga	1		0.04	30–38	33.3	133	126–135	129		1.47
Santie	1		0.07	46–57	52.7	150	110–125	115		4.00
Asachera	3	0.07	0.10	29–49	39.2	108	102–126	114	2.0–2.13	2.09
Asamolgu	2	0.04	0.09	33–97	69.8	91	113–133	121	2.3–4.13	3.30
Agona	3	0.06	0.10	40–76	59.2	91	113–151	137	2.3–2.67	2.49
GR-18	2	0.06	0.10	30–70	51.5	97	80–97	87	2.3–3.27	2.80

The sizes of plots cultivated by female farmers varied from a minimum of 0.02 to a maximum of 0.11 hectares with a mean of 0.06 hectares, confirming the preliminary findings that most of the women farmers cultivate small plots which they manage very well.

Conclusion and the way forward

Two of the identified indigenous rice varieties in the Gore community have high yield potential and compare very well with improved varieties. They also have unique properties that make them preferred by the women farmers.

The researchers therefore intend to continue to work with the women farmers and the community as a whole to promote the sustainable production of these promising rice varieties. This recommendation does not imply that only indigenous rice varieties should be promoted, but that there is a need to give the farmers wider choice and help them perfect their skills in selecting desirable genotypes for their environment.

The issue of availability or adequacy of seed of indigenous rice varieties for planting remains unresolved. From these studies, no formal seed systems exist in the area and farmers plant their fields annually from seed reserved from the previous harvest or by buying from relatives, colleagues, and friends. Such seed exchanges can result in spontaneous crossings between varieties and wild relatives, thus introducing new materials into the existing local gene pool.

There is a need to empower the women to enable them to multiply seed of the indigenous rice varieties to meet their immediate needs and those of other farmers within and outside their community. Support for seed production should also focus on improved storage management, seed health, and purity. Establishment of seed banks in the community where individual farmers' seed or community seed is stored would be particularly helpful. Seed and food fairs based on the indigenous rice varieties have been organized by the farmers and the northern Ghana PLEC team both at Gore and during national meetings of Ghana PLEC, and the potential value of the indigenous rice varieties confirmed.

The *in-situ* conservation of indigenous rice varieties by the women of Gore is thus a success story that needs to be sustained and built upon.

Acknowledgements

The authors are grateful to WAPLEC for funding the studies. The field assistance given by technical staff of the Manga Research Station, headed by Mr Joseph Ali, as well as the activities of Moses Aduko and Simon Asaro at the Kusanaba site, are highly appreciated. The late Charles Anane-Sakyi pioneered this work and the authors wish to dedicate it to his memory.

REFERENCES

Longley, C. and M. Sellu-Jusu, "Farmers management of genetic variability in rice", *ILEIA*, Vol. 15, No. 314, 1999, pp. 16–17.

M'Mukindia, K. L., "Official opening address", in A. Putter, ed., *Safeguarding the Genetic Basis of Africa's Traditional Crops*, proceedings of a CTAIIPGRVICAUIIUNEP Seminar, 5–9 October 1992, Nairobi and Rome: International Plant Genetic Resources Institute, 1994.

10

Vegetables: Traditional ways of managing their diversity for food security in southern Ghana

Essie T. Blay

Introduction

Vegetables constitute a major source of protein, minerals, and vitamins in the diet of resource-poor communities because of the relatively limited access of the people to animal protein sources. They are consumed fresh in sauces or cooked in stews or soups. Often a meal in such communities consists of a starchy staple and a vegetable dish (soup or stew) spiked with comparatively little animal protein. Therefore to complement the staple foods, the rural folk maintain and utilize a wide array of indigenous vegetables, both wild and cultivated.

Vegetables are found in many land-use systems as cultivated crops or volunteers in annual food-crop farms, as part of agroforestry systems, cash crops in mixed cropping systems, monocrops, or as volunteers/wild plants in orchards, uncultivated fallows, or virgin lands.

Recent developments, such as environmental changes, introduction of improved varieties, cash crop production in monocultures, pressure on farmlands, and changes in dietary habits associated with urbanization, are increasingly taking their toll on the array of vegetable germplasm conserved and utilized by the rural populace over several generations (Martin and Ruberte, 1975). If left unchecked, erosion of the indigenous sources of vegetables will compromise the food security of the rural populations and also incapacitate future crop improvement programmes. A cost-effective and sustainable method of ensuring the maintenance of this germplasm is to document traditional knowledge and practices for their management and utilization and to adopt a participatory approach to arrive at the best farmer practices.

This chapter discusses the diversity in common vegetables maintained and used by the rural farm families and the strategies adopted by farmers for their conservation in southern Ghana.

Information discussed was gleaned from a study conducted in three PLEC demonstration sites in southern Ghana, namely Amanase-Whanabenya, Gyamfiase-Adenya, and Sekesua-Osonson (Maps B, C, D, E).

Economic importance of vegetables to the rural household

A wide range of vegetables are collected or primarily produced for the sustenance and food security of the farm family. Only 33 per cent of the farmers produce vegetables for sale. Therefore, vegetables seem to be of minor importance with regard to generation of family income. Only 2.5 per cent of the farmers derive more than 33 per cent of the family income from the sale of vegetables.

The major reason for farmers' choice of vegetables for cultivation is palatability. Others include crop duration, shelf life, high yields, and nutritional value.

Vegetables are gathered from the wild or cultivated on peasant farms as part of a mixed cropping system, or they are grown in home gardens primarily for home consumption. Some annual vegetables are also grown in pure stands, on large separate tracts of land within a multiple cropping system. Under these circumstances the primary objective is income generation.

Commercial vegetable farmers produce comparatively large-scale monocultures, about two hectares in size. Vegetables grown for commerce may be exotic vegetables such as cabbage, lettuce, carrots, onions, cauliflower, sweet peppers, squashes, and cucumbers. Such exotic vegetables are mainly for special local niche markets. In recent times, some Asian vegetables, mostly members of the *Leguminoseae* and *Cucurbitaceae*, have been introduced into the country for production for the export market. In addition to these, a number of local vegetables notably peppers, garden eggs, tomatoes, and okra are also produced on a fairly large scale under truck farming for the urban markets (Sinnadurai, 1992).

Indigenous vegetables in the demonstration sites

The non-exotic or local vegetables found in small farmer holdings in the demonstration site fall into four broad categories, namely:

- the fruiting vegetables comprising the vegetables in the nightshade family (*Solanaceae*), okra and hibiscus (*Malvaceae*), and cucurbits (*Cucurbitaceae*), etc.
- legumes/pods
- leafy vegetables
- bulbs (Table 10.1).

Table 10.1 Major fruiting vegetable cultivars grown in the demonstration sites

Kind	Varieties
Garden egg	Round type, *odwan shua*, white round, *nsorowia*, *twataye*, unknown* (5)
Eggplant	White round, violet type, white long, yellow, unknown* (1)
Tomato	*Navrongo*, lorry tyre, *fa adze begye/kentenma*, *rasta*, *loriano*, *tamale*, processing type, *kumasi*, *denuba*, *daka-daka*, *asere*, power, local/unknown* (4), *Akorle*, *Towenya*, Dakam
Pepper	Curled finger type, *tootoo*, *kpakposhito*, *ogyenma*, *sokwe*, *dalewa* (red, yellow, and white types), *kwadaa yowhi* (bird pepper – several strains), and *toku kwadaa*
Okra	*paa kweku*, *labadi dwarf*, *otrom koli*, *afugbe koli*, 3-month variety, *kpulvor*, *mpempem sono*, *asontem*, tall type, etc.
Sword bean	White, pink, mottled, all in two size classes – small and large
Lima bean	Four-month variety, *onanomobo*, large pink, *akotomo white*, akotomo mottled, long duration types (1–3 years)
Cowpea	Light brown, black eyed, brown mottled
Pigeon pea	Light brown seeded type, white type
Onion/shallot	Shallot (unknown)*
Cucurbits	
Agushie	Unknown*
Snake gourd	Unknown*
Pumpkin	Unknown*

* Variety name unknown.

Fruiting vegetables

The major indigenous fruiting vegetables grown in the southern Ghana PLEC demonstration sites are members of the *Solanaceae*, *Malvaceae*, *and Cucurbitaceae* families.

Solanaceae

*Pepper (*Capsicum *spp.)*

Pepper is one of the most important vegetables in southern Ghana. The pods are ground fresh and prepared into a sauce that is served with a main meal, or used as a spice or condiment in the preparation of stews and soups. Three major species occur locally, namely *Capsicum frutescens*, *Capsicum anuum*, and *Capsicum sinensis*, with over 11 major land varieties.

The *Capsicum frutescens* are hot peppers with mostly elongated, slender fruits that ripen red. These are important commercially and are sold both fresh and dried. A wide variation occurs in fruit sizes and immature colours. *Capsicum annum,* however, is represented by both hot and sweet cultivars. The hot or pungent varieties have more commercial importance than the sweet type. The land varieties of sweet peppers are non-bell types, up to three centimeters in diameter with variable shapes and immature fruit pigmentation ranging from cream/light yellow to various shades of green, all of which ripen red. The *Capsicum sinensis* cultivars, which include *kpakposhito* and *odjengba*, are also very variable in shape and sizes but are usually not elongated. These are called the fragrant peppers on account of their sweet aroma. The immature fruit colour in this species varies from cream through light yellow to various shades of green. The ripe fruit colour is red or yellow. Some of the most pungent fruit types come from this species. Fruits of this species are unsuitable for preservation by drying. Over 11 varieties and strains of peppers are found in the southern zone, comprising both wild and cultivated types.

For home consumption, peppers are normally intercropped on annual food farms as scattered plantings among other vegetables. However, they are also grown in solid blocks in pure stands in a multiple cropping system, or as a monocrop for commercial production. Pepper is an important component of home gardens where a few plants are maintained for year-round supply of fresh pods. Very often one or two plants of an elite selection may be kept close to the kitchen for special protection and easy access.

The garden egg/eggplant complex

The garden egg/eggplant complex comprises eggplant (*Solanum melongena*) – *atropo*, garden egg (*S. gilo*) – *ntrowa*, *S. macrocarpon* – *gboma*, *S. torvum* – *abeduro*, and *nsusua* (*S. nigrum*). Among these the most important on the local market is the garden egg, followed by the eggplant. The fruits of the *Solanum* species are used in the preparation of soups and stews. In the case of the *gboma* (*S. macrocarpon*) the fruits of most varieties are hard and turn corky on maturity. Therefore the leaf is the most important edible part of this species even though fruits of some cultivars are also consumed.

The various species show variability in fruit sizes, immature colour, and taste. The fruits of the eggplant (*S. melongena*) vary in size, shape, and colour. They are rounded or elongated. Immature fruit colour ranges from white through different shades of green to purple. The colours are solid or mottled. Mature fruit colour is usually yellow.

The garden egg (*S. gilo*) fruits on the other hand are mostly white, cream, or various shades of green when immature, but ripen red. They are eaten predominantly in soups and stews both immature and ripe, but the former is preferred.

S. torvum

Members of *S. torvum* show little variation in fruit size or pigmentation. The small, pale-green fruits occur in clusters. The fruits retain their pale-green colour at maturity. The *S. torvum* is mostly self-sown and the fruits are collected from volunteer plants on the farms, around the homestead, or from the wild. *S. torvum* fruits are used in the preparation of soups and stews. They may also be cooked together with palm nuts for the preparation of soup for lactating mothers. Both the fruits and leaves are exploited in the treatment of anaemia.

Tomato

Two species of tomato occur, namely *Lycopersicon esculentum* and *L. pimpinellifolium*, the most important and widespread being the *L. esculentum*. Tomato fruits are widely used in the preparation of sauces, stews, soups, and salads. The *L. esculentum* is grown in home gardens, annual cropping, and monoculture cash cropping systems. Several land varieties exist in addition to imported cultivars (Plate 12). Volunteer or deliberate crosses between the local and exotic cultivars also occur. The local tomato fruits are predominantly scalloped and vary in sizes. The fruit ripens red but a few pink-fruited types occur. The fruits typically have very low total solids content and low acidity, which make them unsuitable for processing. The *L. pimpinellifolium*, locally called *fa adze begye*, is usually grown as single or a few perennial/semi-perennial plants in the home garden or in mixed annual cropping systems. The *L. pimpinellifolium* assumes economic importance during the lean season when tomatoes are scarce.

Malvaceae

Okra (Ablemoschus esculentus)

Two major types of okra (also called "okro") are found, namely an annual short-duration type (the *labadi* dwarf) and semi-perennial types. The perennial types are usually very tall varieties that may grow up to two metres or more at full maturity. The semi-perennial varieties come into bearing about six to 12 months after sowing and continue to fruit for two to three years. The okra fruits vary in pod sizes, shape, colour, and skin texture. The fruits also vary in the extent of sliminess. Okra pods are cooked in stews or soups. The tender leaves of some varieties are also used as spinach or dried, milled, and used in preparing soups and stew (Norman, 1992).

Cucurbitaceae

The major vegetables in the *Cucurbitaceae* family that farmers raise for their consumption include *agushie* (*Cucumeropsis edulis*), *neri* (*Citrullus vulgaris,* syn. *C. lanatus* or *Colocynthis citrullus*), pumpkin (*Cucurbita moschata*), the marrow (*Cucurbita pepo*) and snake gourd (*Trichosanthes cucumerina*), and *chow chow* or *chayote* (*Sechium edulis*).

Agushie and *neri* seeds are used in the preparation of soups and stews. *Neri* is a short-seasoned creeper that matures within 34 months after planting, and is more common in the savanna zone. *Agushie*, however, is a long-season climber that is usually grown on permanent trees. It matures about nine months after planting.

The pumpkin (*Cucurbita pepo*) is grown as edge plantings in home gardens, agroforests, or food farms. The immature fruits and leafy shoots are consumed as vegetables. Other common vegetables in the *Cucurbitaceae* family are the snake gourd and the *chayote* and *Telfaria* sp. One or two plants of snake gourd are grown for home consumption in home gardens supported by live trees. The ripe pulp around the seed is used as tomato paste substitute. The *chayote*, however, is grown in fairly large numbers in pure stands for home consumption and for sale. The immature fruits of *chayote* are used as a garden egg substitute in soups and stews.

Vegetable legumes

The most common legumes observed are the sword bean (*Canavalia ensiformis*), lima bean (*Phaseolus lanatus*), cowpea (*Vigna unguiculata*), pigeon pea (*Cajanus cajan*), and black velvet bean. They occur in food farms or in home gardens. The leaves of some species are used as spinach, fodder, or as a medicine. The beans are commonly boiled and served as a snack, or prepared into stews or in soups, while the beans of some species are prepared into doughnuts and served with porridge or as a snack.

Sword bean (Canavalia ensiformis)

Several land varieties exist, with pole/climbing stems bearing pods of varying sizes that contain white, dark red, purple, or mottled beans of variable sizes, from large beans (2–3 cm long) to small beans about 1.5 cm long. The sword bean is grown in food farms or in home gardens using live trees as stakes. The local varieties are long-season types that last for over nine months. Sword beans are eaten at the immature green stage or mature dry stage. The beans are boiled and served as a snack or side dish, cooked and served whole in soups, or mashed and prepared into soups or stews.

Lima bean (Phaseolus lanatus)

The varieties encountered differ in pod and bean sizes. The colour of beans ranges from white, light brown, red and white, or purple and white mottled, to purple. Again all varieties are pole types. Lima beans are used in much the same way as the sword beans. The leaves are also consumed as spinach.

Cowpea (Vigna unguiculata)

Like the other pulses, the land varieties of cowpeas encountered under forest conditions are all climbers, although creeping and erect types exist in the country. The bean size and colour vary. The common colours are the black-eye types, brown mottled, light brown, and red. Immature pods or beans or mature dry beans are eaten. They are cooked in soups, boiled and prepared into stews, and served with *gari*, a cassava meal, or they are cooked together with rice. The beans are also milled and prepared into doughnuts. The immature leaves are revered as a potherb or spinach.

Pigeon pea (*Cajanus cajan*)

The pigeon pea occurs as semi-perennial shrubs in home gardens or in the food farms. Only a few plants are kept, primarily for home consumption. Both the mature and immature seeds are cooked and served as a snack, in soups, and in stews. The leaves are said to have medicinal uses and are also used as fodder. It is primarily used as a food security crop. The mature beans are very hard and take a long time to cook. On the other hand the hard seed coat aids in preservation of the beans.

Black velvet beans, adua apea

The black velvet bean is a very hairy plant, which causes skin irritation. The crop used to be consumed as a hunger crop. A non-hairy, non-irritating strain occurs, which is boiled and served in stews or soups.

Leafy vegetables

About 20 major species of leafy vegetables are commonly used in the demonstration site. They comprise both wild and cultivated plants: annuals or perennials, herbs, shrubs, and trees. Leafy vegetables abound during the rainy season as wild or volunteer plants around the homestead, in home gardens, crop farms, in bushes, or on wastelands. The most common examples are water leaf (*Talinum triangulare*), stinging nettle (*Fleurya estuans*), amaranthus (*Amaranthus hybridus*), cocoyam leaf (*Xanthosoma saggittifolium*), and *Mormondica charantia*, locally known as *nyanyina*.

Other leafy vegetables are specifically cultivated in home gardens or crop farms during the cropping season. Notable among these are *gboma* (*Solanum macrocarpon*), *ademe* (*Corchorus capsularis*), bitter leaf (*Venonia amygdalina*), and the *chaya*.

The leaves of certain crop plants grown primarily for other plant parts such as some roots and tubers are also harvested and consumed as vegetables. These include cocoyam, sweet potato (*Ipomea batatas*), cassava (*Manihot esculenta*), and taro (*Colocasia esculenta*). Others are cowpea, okra, garden egg, lima bean, and pumpkin.

During the dry season, leafy vegetables obtained from home gardens and wetlands are supplemented with the tender leaves of trees such as the silk cotton (*Ceiba pentandra*), baobab (*Adansonia digitata*), and cocoa (*Theobroma cacao*). The tender leaves of papaya (*Carica papaya*) are also used as a vegetable, but preferably during the rainy season. Other lesser-known food/spice leaves include *wawa* (*Triplochitan scleroxylon*) leaf, mint and wild mint (*Ocimum basilicum*) leaves, *nyanyina* (*Mormordica charantia*) leaves, leaves of *okosow* (*Sterculia* sp.), and wild lettuce (*Lactuca taraxifollium*).

Some important leafy vegetables in the demonstration sites

Cocoyam leaves (*Xanthosoma saggitifolium*)

Cocoyam leaves are the most popular leafy vegetable in Ghana. It may appear as a volunteer crop after land clearing, especially in areas with a previous history of cocoa production. It is also cultivated from split corms or from cormels. Tillers produced by leaving the cormels of mature cocoyams in the ground to sprout into thick clumps of suckers are divided for planting during the rainy season. Cocoyams are grown either in the home garden, annual crop farm, or both. The leaves are harvested until the environment is too dry to support new leaf production, and the old leaves become too coarse or too dry for consumption. The tops are then allowed to die off and to resume growth at the beginning of the rains.

Taro leaves (*Colocasia esculenta*)

Taro leaves are used in the same way as cocoyam leaves, although they are less popular than the latter. Taros are found in marshy areas in crop farms or growing wild on wetlands. In home gardens and around the homestead, they are cultivated in areas where wastewater runs. Taro leaves are therefore more frequently available year round than cocoyam leaves (Irvine, 1979).

Water leaf (*Talinum triangulare*)

Water leaf grows wild during the rainy season. They are available for picking in large quantities throughout that season and beyond, where water is available. There is a market for the species. Therefore the leaves are sometimes picked for sale. The talinum leaves are not usually preserved. However, one farmer reported that cutting them into pieces might preserve the leaves, sprinkling a small quantity of wood ash on it to reduce the slime and sun-drying.

Cassava leaves (*Manihot esculenta*)

Young leaves of the cassava are used as vegetables, especially in Amanase-Whanabenya. Cassava leaves are used as a single leafy vegetable or are mixed with several other leaves including papaya to prepare a nine-leaf soup. However, cassava is mainly planted for its tuber that is processed into *gari* or pounded into *fufu*. The utilization of the leaves as a vegetable is only secondary. The crop is maintained by continuous cropping using stem cuttings from the harvested crop. Also, new introductions are continuously made.

Sweet potato (*Ipomoea batatas*)

The sweet potato is grown primarily in annual food farms, or in home gardens. They are propagated on mounds or beds from vine cuttings. The use of sweet potato leaves as spinach is not widespread.

Gboma (*Solanum macrocarpon*)

Gboma species, like *Ademe* (*Corchorus olitorius*) and *Sorbui*, are popular leafy vegetables in the Volta region, home of the Ewe-speaking people of Ghana. The three species are found primarily in settlements of Ewe migrants in the demonstration sites. The culture of *gboma* is similar to that of garden eggs. Both are normally nursed and transplanted. The leaves are cut up and used in the preparation of soups. *Gboma* is cultivated mainly in home gardens, but sometimes it occurs in crop farms. Where water is available, it is maintained as a perennial or replanted from seeds that are saved from one planting season to the next.

Ademe/bush okra (*Corchorus olitorius*)

Ademe is propagated from seeds. It occurs primarily in settlements of Ewe migrants. In one non-Ewe settlement where the crop was found, the respondent had stayed in northern Ghana for a long time and had acquired the taste from there. The crop is maintained from seeds that are saved from mature crops from one season to the next.

Amaranthus or aleefu (*Amaranthus hybridus*)

Amaranthus is a popular vegetable savoured by the northern communities. It is easily established by broadcasting the dry seeds. It also grows wild and matures four to six weeks after sowing. Amaranthus is harvested by uprooting the whole plant or by topping and used in the preparation of soups or stews. The crop normally regenerates at the beginning of the rainy season from fallen seeds. In a few cases, the farmers save the panicles containing the seeds for future planting.

Stinging nettle/honhon (*Fleurya aestuans*)

The stinging nettle grows wild in plantations, agroforestries, and fallows. The leaves are reported to be excellent for pregnant women. They reportedly aid proper foetus development and prevent anaemia. This leaf is best known to the Twi-speaking ethnic group.

Nyanyina (*Mormordica charantia*)

Nyanyina is a climbing plant. It grows wild, especially on newly opened forests. The leaves are used in soups or stews. They are mostly collected from the wild. The seeds may be collected and dried for future planting. The leaves are also used for the treatment of snake bites and hypertension.

Leaves of pawpaw (Carica papaya)

Young pawpaw leaves are mixed with other leaves for the preparation of stews. It is used mainly during the rainy season, when fresh growth occurs.

Wild mint (Ocimum basilicum)

The fresh leaves of wild mint are mixed with other leafy vegetables for the preparation of stews.

Wawa leaves (Triplochiton scleroxylon)

The leaves of the *wawa* tree are mixed with cocoyam leaves for the preparation of stews or soups. They are harvested from the wild or sometimes from around the homestead.

Okra (Ablemoschus esculentus)

Okra is usually grown for its slimy pods. It is grown in annual cropping systems, home gardens, or in the homestead. In the homestead, usually the semi-perennial types are found. The leaves are sometimes harvested for the preparation of soups. Saving seeds from one planting season to the next maintains the crop. In the case of some perennial varieties, seeds that dehisce from pods may regenerate into new plants. Some of the tall, semi-perennial varieties are coppiced to extend the harvest duration.

Cowpea leaves (Vigna unguiculata)

The cowpea is a highly nutritious vegetable, but is better known by communities in northern Ghana. Although cowpea was recorded in all the three demonstration sites, nowhere was it mentioned as a leafy vegetable. The crop is mainly grown for its beans.

Baobab (Adansonia digitata)

The baobab tree is known locally as *kuka*. The tree is grown in the farmstead or on the farms, and the leaves are harvested for the preparation of stews and soups.

Management of vegetable crop diversity

Sources of planting materials of vegetables

Farmers depend heavily on their own seed sources as well as seeds from the local markets. While 77 per cent of the farmers raise their vegetables from seeds saved from their crops, 89.7 per cent purchase some of their seeds from the local market. Seeds are also frequently obtained from neighbours (67 per cent). Other systems such as exchanging germplasm or even working in exchange for new seed are used.

In the case of semi-wild vegetables such as *Solanum torvum*, bird pepper (*Capsicum anuum*), cocoyam (*Xanthosoma saggitifolium*), wild lettuce (*Talinum triangulare*), and a number of leafy vegetables, volunteer plants feature prominently (77 per cent) as a source of planting material. Other minor sources of vegetable seeds include agricultural shops, research stations, and farmers' association farms.

The diversity of vegetables that are cultivated is in a constant flux. Nearly 60 per cent of the farmers introduce some commercial varieties into their farm. About 50 per cent of them drop some vegetables permanently or during some planting seasons for several reasons, including ease of incorporation into mixed cropping systems, resistance to drought and excessive rains, and market demand. New vegetable species/varieties are also added.

Maintenance of vegetables

Some vegetables grown in the home gardens are strictly annuals while others are maintained in the field for over a year. Almost all farmers maintain, in the field for almost a year, some vegetables, notably aubergines (*Solanum melongena*), okra (*Ablemoschus esculenta*), bush okra (*Corchorus olitorius*), *agushie* (*Cucumeropsis edulis*), and the perennial leafy vegetables such as the *chaya*, cocoyam, and bitter leaf (*Vernonia amygdalina*) as well as trees with edible leaves such as silk cotton and baobab.

Special strategies are adopted to ensure the survival of these vegetables in the field during the dry season. They include the following:

- weeding around the crop during the dry season
- watering where water is available
- weeding around the plants before the onset of the dry season
- leaving plants in the bush during the dry season
- planting cassava around the plants to provide shade
- planting in the home garden where it can be watered during the dry season
- mulching
- setting traps to ward off animals.

A few (one to three) plants of élite or perennial varieties of vegetables that are intensively used, such as okra, peppers, *Solanum torvum*, eggplant, tomato, and taro are often grown very close to the kitchen for intensive care and ease of accessibility (Plate 12).

Maintenance of vegetable seeds

Maintenance of vegetable seeds is one of the most important ways of ensuring the continuous propagation of non-recalcitrant vegetables for food security.

In the case of obligatory annual vegetables, seeds are either extracted or stored in dried pods/dried fruits or both. Seeds are stored for three to 36 months, with 12 months as the modal storage period, followed by three months and six months. A 12-month storage period implies saving seeds from one major planting season to the next, while three-month storage applies to seeds that are stored from the minor season harvest for the major season.

Farmers usually keep vegetable seeds wrapped in rags/paper or in corked bottles. Fruits of pepper and garden eggs are sometimes strung on sticks and dried. The dried fruits are subsequently stored in kitchens, rooms, cupboards, and wooden boxes or in any cool dry place. Some vegetable planting materials (dried seeds and pods) are also stored in bags or baskets that are hung on walls until needed. A few farmers have adopted a new seed storage technology whereby well-dried vegetable seeds are put in tightly capped bottles and stored in a well-covered hole under shade in the farm or home garden. At the beginning of the planting season, farmers determine seed viability using the floatation test, eye inspection for colour change or insect damage, or by planting a sample for a germination test. About 10 per cent of farmers do not carry out any check for seed viability.

Other indigenous seed-testing practices include throwing a sample into fire and identifying viable seeds from popping sounds or biting into the seed to check for colour and texture to determine viability.

Maintenance and conservation of diversity in vegetables for food security

Generally, the strategies adopted for vegetables are similar to the methods used for conservation in yams in southern Ghana (see Chapter 7). In addition to the strategies outlined for yams, the following methods are also used for maintenance of vegetable germplasm.

- Perennial and semi-perennial vegetables such as cocoyams and some okra and pepper varieties are maintained in the farms by weeding around them just before the rains and mulching at the onset of the dry season.
- Seed storage is rigorously pursued. Seeds are mostly stored by wrapping them in cloth or paper and hanging or pushing them under roofing members in the kitchen where smoke from the kitchen fire fumigates and keeps them dry to help to discourage pests and diseases.
- Okra seeds are stored in pods or shelled and tied in a cloth and stored over the kitchen smoke, or in corked bottles that are stored in a cool, dry place, usually in a room.
- Plants of precious heirloom varieties are maintained close to the kitchen for special care.
- Maintenance of highly biodiverse home gardens where germplasm is kept also helps to conserve the vegetable germplasm.

- Viability of stored seeds is closely monitored using indigenous technologies.
- Strategic exploitation of wild annual vegetables and leaves of tree species help to supplement the cultivated types.

Conclusion

The vegetable germplasm, especially of the fruit vegetables and legumes, is very scant compared to the roots and tubers, especially yams. Even though as many as 20 different leafy vegetables exist in the three demonstration sites in southern Ghana, the utilization of many vegetables is dictated by ethnic origin.

There is a need to promote the utilization of indigenous vegetables, notably the leafy vegetables. At the village level the use of *S. torvum* is ubiquitous owing to its perceived health benefits and, possibly, its drought tolerance.

A number of factors, both natural and man-made, now threaten the continued maintenance of the diversity. Fortunately, the farmers are keen on conserving the heirloom varieties. With a little support, they can be counted on to maintain the diversity for posterity.

Among the knowledge gaps identified was the need to develop the post-harvest and processing aspects of both fruiting and leafy vegetables. This problem should be addressed to ensure year-round supply of vegetables.

REFERENCES

Irvine, F. R., *West African Crops*, London: Oxford University Press, 1979.

Martin, F. W. and R. M. Ruberte, *Edible Leaves of the Tropics*, Mayaguez, Puerto Rico: Antillian College Press, 1975.

Norman, J. C., *Tropical Vegetable Crop*, Devon: Arthur H. Stockwell, 1992.

Sinnadurai, S., *Vegetable Cultivation*, Accra: Asempa Publishers, 1992.

11

The *proka* mulching and no-burn system: A case study of Tano-Odumasi and Jachie

Charles Quansah and William Oduro

Introduction

Traditionally smallholder farmers in Ghana burn the vegetational debris after clearing virgin or fallowed land. They view burning as the most convenient, efficient, and economic way of getting rid of the large amounts of woody biomass which often characterizes forest clearance and constrains cultivation. The fire destroys weed seeds, pathogens, pests, and snakes, and the ash produced improves soil fertility.

However, burning off vegetation, particularly by indiscriminate use of fire for land preparation, contributes significantly to deforestation, destruction of biodiversity, and depletion of a primary source of soil nutrients and the natural sink for carbon dioxide. The clearing of forests also impacts adversely on the habitat and migratory routes of wildlife, and increases insolation and soil temperature, which adversely affects the activities of useful soil microbes. Large-scale clearance of forest by fire contributes significantly to climate change not only at the local level as observed by the aged farmers at the PLEC sites, but also at the global level through increased levels of carbon dioxide in the atmosphere. The latter, in turn, increases the greenhouse effect. Additionally, forest functions such as watershed protection, stormflow stabilization, runoff control, soil erosion prevention, and environmental amelioration are adversely affected (Quansah *et al.*, 2002).

The adverse effects of the slash-and-burn system override its transient benefits. Because of this, the slash-and-burn system has not been able to sustain soil fertility and productivity and biodiversity conservation under intensive cropping systems.

To overcome this problem with a view to enhancing the livelihood of smallholder farmers requires the development of participatory and sustainable models of land and biodiversity management based on farmers' technologies and knowledge within their farming systems. Recognizing this approach as one of the major objectives of PLEC work in Ghana, a search was made to identify sustainable traditional land management practices within the PLEC demonstration sites, Tano-Odumasi and Jachie (Map A). The search revealed the existence of a slash-and-no-burn system, which, in the Akan-speaking areas, is referred to as *proka*, or *oprowka*, its dialectal variation. It literally translates "add to by rotting". Because it favours biodiversity conservation and agricultural production but is dying out, PLEC has directed its attention to reviving and encouraging its practice through demonstrations by expert farmers.

This chapter describes the *proka* system with reference to the PLEC demonstration sites at Tano-Odumasi and Jachie in the moist semi-deciduous forest zone of central Ghana (Map B).

Methodology

The methods used for the study consisted of desk work and fieldwork. The desk work involved a review of relevant literature on the subject from as many sources as possible and accessing the general survey data accumulated in the PLEC database. Where data on aspects of the subject matter are lacking for the PLEC sites, other relevant data from central Ghana are used and duly acknowledged.

The fieldwork involved iterative dialogue sessions and interactions between a multidisciplinary team of researchers and collaborating PLEC farmers under the UNU/PLEC-Ghana project. Emphasis was laid on focused group discussions, informal interviews, particularly with expert farmers and the aged, and visits to and observations at farmers' fields.

Soil samples were taken from various land-use stage sites at Tano-Odumasi to a depth of 30 cm at 15 cm increments for routine chemical, physical, and biological analysis. The land-use stages studied comprised agroforestry (*proka* managed), annual cropping (*proka* managed), annual cropping (slash-and-burn), and native forest (sacred grove).

Discussion of findings

Proka *defined*

Proka is a land management practice whereby the vegetation cleared in the course of land preparation for farming is left in place without burning.

Crops are subsequently planted through the plant biomass, which forms a mat of mulch and decomposes *in situ* to add organic matter and nutrients to the soil for the benefit of the crops. Using the definitions of agrobiodiversity terminology recognized by the PLEC Biodiversity Advisory Group (BAG; Zarin, Huijin, and Enu-Kwesi, 1999) *proka* is an element of management diversity. It is neither a land-use stage nor a field type.

Proka *in practice*

In the *proka* traditional method of land preparation, the vegetation is cleared in the dry season, December/January, and allowed to dry and commence decomposition till about February/early March. The biomass is subsequently chopped down using the cutlass to facilitate planting of crops at the onset of the rains. Planting is done after the rains are well established. The type of land clearing is determined by the land-use stage and the field type envisaged.

Where and when is proka *practised*

At the PLEC sites in central Ghana, *proka* is practised in land-use stages comprising agroforests, orchards, and annual cropping. It is also practised when the farmer is late in preparing the land to meet the onset of the rains.

Agroforest

In the agroforest land-use stage where the envisaged field type consists of a mixture of staple crops and trees, land clearing comprises clearing of herbaceous plants, shrubs, and small trees using the cutlass. Big trees are felled either by axe and/or by fire. Some trees are deliberately left *in situ* with the density consciously managed to permit adequate insolation for the proper growth of the crop mix to be planted. *Proka*, as described above, is the main land preparation method.

In both PLEC sites, a typical agroforest consists of a field type made up randomly of a planted mix of plantain, some fruits trees, and, in some cases, maize. Subsequent land management consists of weeding with the cutlass and/or the hoe, and leaving the vegetational debris in place to rot. In some cases, vegetables are planted at spots where slashed residues are heaped around tree stumps or termite mounds and allowed to rot, or are burnt (Quansah *et al.*, 2001).

Orchard

Proka as practised in the semi-deciduous forest zone of Ghana originated mainly as the dominant land preparation method for establishing cocoa plantations, which are examples of orchard. Land clearing for cocoa plantations is similar to that described above for agroforest, except that the tree density for the cocoa plantation is greater since the latter crop requires shade.

In its early establishment, the field type consists of cocoa seeded randomly at stake and intercropped with plantain, cocoyam, and cassava, which supplement the shade provided by the trees for the young cocoa seedlings. The landuse at the early stages and during the growing phase may be more aptly described as agroforest than orchard. However, when the cocoa canopy closes the intercrops can no longer survive, and the field type evolves into a cocoa-dominated plantation interspersed with the original trees left in place during land preparation. This evolution brings into question whether this final land-use stage should be called agroforest or orchard within the context of the BAG definitions (Zarin, Huijin, and Enu-Kwesi, 1999).

At the closure of the canopy, the floor of the cocoa plantation is almost always covered by a mat of mulch made up mainly of senesced cocoa leaves. This, in addition to the reduced insolation by the closed canopy, effectively controls the growth of weeds. The management of the field at the early stages of establishment consists of frequent weeding, which turns into occasional brushing once the cocoa canopy is closed. The weed biomass is left to augment the mulch on the floor, which decomposes to enhance soil organic matter content.

The current practice of *proka* in this land-use stage, as in others, consists of an initial land clearing followed by the burning of the biomass. This reduces the tree density and permits the recent introduction of the light-demanding maize as an intercrop in addition to food crops. The field is, therefore, deprived of the benefits of the biomass at the establishment stage. With the intervention of cocoa extension by the Ministry of Food and Agriculture (MOFA), line planting of raised cocoa seedlings is increasingly replacing seeding at stake. This, according to the PLEC collaborating farmers, reduces wastage of planting material and facilitates weeding and systematic spraying to control pests and diseases in the cocoa plantation.

Annual cropping

Annual cropping is a land-use stage often practised in fields with short fallow periods. At the PLEC sites in central Ghana, the fallow vegetation often consists of *Chromolaena odorata*, *Panicum maximum*, *Aspilia africana*, *Sida acuta*, *Griffonia simplicifolia*, other herbaceous plants, and shrubs.

Generally land preparation consists of total land clearance and burning of the biomass followed by planting on the flat, mounds, or ridges. However, because of the short fallow period, biomass regrowth is limited. This has made it possible for a few of the PLEC farmers to practise *proka* in this land-use stage with maize/cassava intercrop as the sole field type. Land management after the initial establishment consists of weeding as and when necessary, leaving the weed biomass in place to rot. In most cases no mineral or organic fertilizer is applied. The maize is harvested within three months. Thereafter the cassava remains as the field type till its harvest.

Other field types under this land-use stage, but without *proka*, include monocrops/maize and vegetables (tomato, okra, garden egg, pepper, and cabbage). Among all the field types, maize/cassava intercrop is the most common. Monocrops of vegetables are popular among younger farmers who practise market gardening for the nearby urban market in the city of Kumasi. The vegetables are often cultivated in valley bottoms with or without the application of mineral or organic fertilizers. Supplementary irrigation cultivation of vegetables is often characterized by indiscriminate use of pesticides.

Biodiversity and land-use stage

Biodiversity values for various land-use stages at Tano-Odumasi comprising diversity and heterogeneity indices for three plot sizes are presented in Table 11.1. The values show biodiversity indices to be higher in the *proka*-managed agroforest than the non-*proka* annual cropping system. This implies that the *proka* management system promotes biodiversity conservation. The values for the *proka*-managed agroforest compare favourably with those of the native forest (sacred grove) used as the control for the highest biodiversity conservation land-use stage.

Table 11.1 Diversity and heterogeneity indices for various land-use forms (for three plot sizes) in Tano-Odumasi demonstration site

Plot area (m^2)	Land-use form	Diversity and heterogeneity indices					
		Margalef	Menhinink	Gleason	Shannon	Simpson	Brillouin
400	1	0	0	0	0	0	0
400	2	3.51	2.77	3.90	2.20	0.05	1.54
400	4	2.92	2.41	3.34	1.97	0.73	1.37
400	6	2.16	2.41	2.89	1.39	0	0
400	7	0	0	0	0	0	0
400	8	8.33	4.12	8.55	3.26	0.05	2.75
400	9	2.02	1.73	2.41	1.66	0.14	1.22
400	12	0	0	0	0	0	0
25	1	4.77	0.85	4.90	1.98	0.23	1.95
25	2	6.08	1.28	6.22	2.17	0.22	2.11
25	4	5.99	1.37	6.14	2.04	0.27	1.97
25	6	4.31	1.22	4.47	2.05	0.19	1.97
25	7	4.59	2.38	4.82	2.46	0.12	2.13
25	8	7.90	2.35	8.06	3.22	0.06	3.04
25	9	6.82	1.79	6.98	2.06	0.23	3.61
25	12	5.55	1.54	5.71	1.75	0.32	1.66
1	1	3.93	0.51	4.05	1.33	0.61	1.12
1	2	6.41	1.10	6.54	2.39	0.14	2.35
1	4	7.62	2.73	7.80	3.04	0.01	2.80
1	6	6.52	1.54	6.67	2.61	0.14	2.52
1	7	2.87	2.09	3.19	2.11	0.10	2.04
1	8	7.11	2.24	7.27	2.55	0.14	2.39
1	9	5.41	1.42	5.70	2.42	0.16	2.33
1	12	6.03	1.40	6.17	2.44	0.16	2.36

Source: PLEC field survey

Notes: Land-use codes:

1. Annual cropping; 2. Agroforestry; 3. Grass-dominated fallow;
4. Shrub-dominated fallow; 6. Orchard plantation; 7. Orchard trees; 8. Native forest (sacred grove); 9. House garden; 12. *Proka* (agroforestry).

Land-use stage versus soil micro-organisms and fauna

Soil monoliths (25 × 25 × 30 cm) were sampled from random points about five metres apart for three land-use stages, namely native forest (sacred grove), *proka*-managed agroforest, and slash-and-burn annual cropping fields.

Taxonomic units of soil fauna were ants, arachnids (spiders), gastropods (snails), and *Glyllidae* (crickets). Far fewer soil fauna were found in the slash-and-burn soil than the *proka* and sacred grove. The most numerous soil fauna identified were ants. These occurred in all the three monoliths under the three land-use stages.

Over 73 per cent of the ant units were identified in *proka*, 15 per cent in the slash-and-burn annual cropping, and 11 per cent in the sacred grove. Only a single spider was found. It occurred under the grove. Similarly, the three snails and the two earthworms encountered were identified in only the *proka* soil, while the only three crickets that were found occurred in the grove. The most numerous colonies of fungi (33×10^6) and bacteria (36×10^6) occurred in the *proka*, followed by the slash-and-burn soil (30×10^6 and 21×10^6) and sacred grove (9×10^6 and 2×10^6). The *proka*-managed agroforest was thus superior in maintaining higher diversity of soil micro-organisms.

Soil physical properties versus land-use stages at Tano-Odumasi

The physical properties under the various land-use stages were assessed by measuring bulk density and total porosity. Bulk density (Table 11.2) in all the land-use stages was lower in the topsoil (0–15 cm) than in the subsoil (15–30 cm). Consequently total porosity was higher in the topsoil. This is mainly due to the higher organic matter (Table 11.3) recorded in the topsoil. A comparison of the values for the slash-and-burn and *proka* annual cropping systems indicated a lower bulk density and higher total porosity for the latter management system. In all cases, the sacred grove had the least bulk density and the highest total porosity. When averaged over the 30 cm depth, the bulk density ranked in an increasing order of slash-and-burn > agroforest > *proka* annual cropping > sacred grove with their respective values of 1.47, 1.39, 1.33, and 1.20 g/cm^{-3}. The converse, sacred grove > *proka* annual cropping > agroforest > slash-and-burn was true for total porosity with their respective values being 54.72, 50.08, 47.45, and 44.34 per cent. The implications for these values are that those land-use stages with lower bulk densities and higher total porosities have the greater potential to enhance moisture conservation through increased infiltration and low runoff rates. This, together with the mulch cover, would be more potentially effective in controlling soil erosion by water.

Table 11.2 Soil physical properties under various land-use stages at Tano-Odumasi

Land-use stage	Field type	Soil depth (cm)	Bulk density (g/cm^{-3})	Porosity (%)
Agroforest (*proka* managed) JC	Plantain, cocoyam. cassava, trees, oil-palm, fruit trees (mango, pear)	0–15	1.31	50.41
		15–30	1.47	44.49
		Mean	1.39	44.45
Annual cropping (*proka* managed) Barnie	Maize/cassava intercrop	0–15	1.19	55.09
		15–30	1.46	45.06
		Mean	1.33	50.08
Annual cropping (slash-and-burn) JC	Maize/cassava intercrop	0–15	1.29	51.00
		15–30	1.65	37.67
		Mean	1.47	44.34
Native forest (sacred grove)		0–15	0.92	65.42
		15–30	1.48	44.02
		Mean	1.20	54.72

Source: Based on data from PLEC field survey

Soil chemical properties versus various land-use stages at Tano-Odumasi

The results of the chemical analyses (Table 11.3) indicated all nutrients in the topsoil (0–15 cm), except phosphorus and potassium in the *proka* annual cropping, to be generally higher in the *proka*-managed land-use stages. However, the *proka* annual cropping recorded the lowest levels of phosphorus and potassium. The higher levels of phosphorus and potassium on the freshly prepared field of the slash-and-burn annual cropping may be due to the ash from the burnt biomass and an earlier application of NPK fertilizer to a tomato crop on that field.

In the sacred grove there was a higher content of nutrients in the subsoil (15–30 cm) than in the top 0–15 cm.

Residual effect of slash-and-burn and no-burn on soil chemical properties

Slash-and-burn and slash-and-no-burn have significant residual effects on the chemical properties of soils. This is demonstrated by a study of farmers' fields at Nkawie in central Ghana (Tables 11.4 and 11.5; Quansah *et al.*, 1998; Amoakohene, 1999). The land-use stage and the field type are annual cropping and maize/cassava intercrop respectively. The mulch was a mixture of *Panicum*

Table 11.3 Soil chemical proprieties under various land-use stages at Tano-Odumasi

Land-use stage	Field type	Soil depth (cm)	PH 1:25 (H_2O)	C %	Organic matter %	N %	P (mg/kg^{-1})	K ($cmol/kg^{-1}$)	Ca ($cmol/kg^{-1}$)	Mg ($cmol/kg^{-1}$)
Agroforest (*proka* managed) JC	Plantain, cocoyam. cassava, trees, oil-palm, fruit trees (mango, pear)	0–15	7.19	0.75	1.29	0.084	27.00	1.660	1.20	4.80
		15–30	7.24	0.21	0.36	0.028	25.25	0.072	1.60	2.40
		Mean	7.22	0.48	0.83	0.056	26.13	0.866	1.40	3.60
Annual cropping (*proka* managed)	Maize/ cassava intercrop	0–15	7.29	1.26	2.17	0.133	6.50	0.020	8.00	0.50
		15–30	7.25	0.93	1.60	0.098	5.00	0.020	6.00	0.70
		Mean	7.27	1.10	1.89	0.116	5.75	0.020	7.00	3.48
Annual cropping (slash-and-burn) JC	Maize/ cassava intercrop	0–15	7.18	0.64	1.11	0.070	25.50	0.165	1.60	4.00
		15–30	7.14	0.30	0.52	0.058	20.00	0.093	1.40	2.80
		Mean	7.16	0.47	0.82	0.063	22.50	0.129	1.50	3.40
Native forest (sacred grove)		0–15	7.22	0.34	0.59	0.014	36.50	0.067	1.40	2.20
		15–30	6.85	1.45	2.51	0.196	25.50	0.165	1.00	4.00
		Mean	7.04	0.90	1.55	0.105	31.00	0.116	2.70	3.10

Source: Based on PLEC survey

Table 11.4 Residual effect of treatments and initial soil value (0–15 cm depth)

Treatment	% N	% OM	% Org. C	PH	P (mg/kg)	K (cmol/kg)	Na (cmol/kg)	Mg (cmol/kg)	Ca (cmol/kg)
Initial	0.09	1.43	0.83	6.80	6.00	0.13	0.08	3.20	6.52
T_0	0.07	0.89	0.51	6.60	4.22	0.10	0.07	1.84	5.24
T_1	0.11	1.43	0.83	6.50	8.18	0.14	0.11	1.60	7.68
T_2	0.13	2.28	1.32	7.20	16.02	0.21	0.11	3.28	9.88
T_3	0.13	2.21	1.28	7.10	14.68	0.20	0.11	2.72	9.16
LSD (1%)	0.04	0.62	0.51	0.55	4.14	0.04	0.02	–	3.48
LSD (5%)	0.02	0.45	0.26	0.39	3.00	0.03	0.014	1.65	2.52
CV (%)	39.84	47.02	45.37	7.82	55.51	30.23	36.69	58.18	31.85

Source: Amoakohene (1999)

T_0 Burnt field + no external input.
T_1 Residues lest + no external input.
T_2 Residues left + 4 t/ha^{-1} poultry manure (PM).
T_3 Residues left + 4 t/ha^{-1} PM + low NPK fertilizer (30-20-20 kg/ha^{-1}).

Table 11.5 Residual effect of treatments and initial soil value (15–30 cm depth)

Treatment	% N	% OM	% Org. C	PH	P (mg/kg)	K (cmol/kg)	Na (cmol/kg)	Mg (cmol/kg)	Ca (cmol/kg)
Initial	0.04	0.65	0.38	6.80	6.00	0.08	0.08	2.44	5.16
T_0	0.03	0.48	0.28	6.60	3.84	0.08	0.07	1.60	4.00
T_1	0.04	0.63	0.36	6.40	5.84	0.09	0.10	2.44	3.92
T_2	0.06	0.73	0.42	7.20	9.18	0.10	0.11	1.40	6.80
T_3	0.06	0.72	0.42	7.10	12.38	0.11	0.11	1.08	6.72
LSD (1%)	0.02	–	–	–	–	–	0.03	–	–
LSD (5%)	0.01	0.21	0.12	0.61	5.04	0.03	0.02	1.66	2.46
CV (%)	51.71	46.55	46.66	10.57	65.28	32.40	36.94	69.59	38.17

Source: Amoakohene (1999)

maximum and *Chromolaena odorata* applied annually at the rate of about five tonnes per hectare. The poultry manure was layer manure.

The results (Table 11.4) showed burning (T_0) to have adverse effects on soil chemical properties. Compared with the initial soil nutrient values, burning depleted all the nutrients measured, namely N, P, K, Na, Mg, Ca, and depressed soil pH. The reduction of organic matter by burning is of grave concern considering that soil organic matter serves as a source of nutrients, improves soil physical conditions, increases the CEC (cation exchange capacity) of the soil, buffers the soil against pH fluctuation, increases the efficiency of nutrient use, particularly applied fertilizers, and generally enhances the productivity of the soil. As burning is an integral part of the farming systems of smallholder farmers, it is not surprising that such farms are characterized by low soil fertility and declining yield.

The results demonstrated clearly that leaving crop or plant residues (T_1) on farmlands makes more sense than burning them. The slash-and-no-burn without any external input (T_1) maintained the fertility of the soil. Soil nutrients at the surface were either higher or the same under the residue than the initial content, although the differences were not significant. The residue protected the soil surface against any possible losses through erosion. The decomposition of the residue over the years may have slowly released some nutrients to the soil.

When the residue was combined with poultry manure (T_2), the initial soil nutrients in both the topsoil and subsoil were significantly increased. The increase was several orders of magnitude greater than that of the residue alone. The increase was ascribed to the additional release of nutrients from the decomposing poultry manure. The buffering capacity of increased organic matter became evident under T_2 treatment. The initial pH of 6.8 was increased to 7.2. The neutral conditions thus created could enhance the release and efficient use of nutrients. It is also remarkable to note that under the poultry manure and residue (T_2), soil nutrients were maintained at the same level as that under poultry manure + residue + mineral fertilizer (T_3). Even in some cases, such as for organic matter, P, Mg, and Ca, higher levels were recorded under T_2 than T_3. This has important implications for the smallholder farmer who, in the wake of increased prices of fertilizers due to the total withdrawal of government subsidy, can no longer afford to buy mineral fertilizers. Where poultry manure is available, as at Tano-Odumasi, farmers can use it for sustained high crop yield, bearing in mind to apply supplementary mineral fertilizers as and when it becomes necessary. The cost of the farmer's soil fertility replenishment can therefore be significantly reduced.

It must, however, be pointed out that the addition of mineral fertilizers to poultry manure + residue (T_3) also significantly increased the initial soil nutrient status in spite of uptake by the test crops, maize/cassava. Organic matter content was also increased, as well as the soil pH. It was expected that the addition of mineral fertilizer (T_3) would give a higher increase in soil nutrients than T_2. However, as indicated by Kang and Balasubramanian (1990) the presence of organic matter through the addition of the poultry manure could have enhanced the efficiency of uptake of the readily available nutrients in the fertilizers by the test crops. In such a situation, the level of nutrients in the soil may either be the same or less than what was recorded under T_2. However, efficient uptake of nutrients may result in higher crop yields in T_3 than T_2 (Table 11.8).

Mulching rates versus nutrient, runoff, and soil losses

For lack of data at the PLEC sites, the data in Table 11.6 (Quansah *et al.*, 2000) are used to demonstrate the effect of *proka* on soil, runoff, and nutrient losses. The experiments were carried out on runoff plots at the University of Science and Technology, Kumasi, which is about 20 km from Tano-Odumasi. The soil was a ferric acrisol with a slope of 3.5 per cent. The mulching material consisted of dry guinea grass, *Panicum maximum*. The land-use stage was annual cropping with maize as the field type.

The results showed mulching to reduce both soil loss and runoff in a decreasing order of bare > no mulch > 2 t/ha mulch > 4 t/ha mulch > 6 t/ha mulch. As the mulching rate increased, the soil loss and runoff decreased. Similarly, organic matter and nutrient losses declined with an increasing rate of mulching. The maintenance of mulch on the soil surface has been found to protect the soil against raindrop impact, impede the flow of runoff, reduce nutrient losses, soil detachment, and dispersion, and maintain high soil infiltration rate (Lal, 1976; Roose, 1976).

The high losses of organic matter are of particular concern because it plays a significant role in the nutrient and water-holding capacities of soils.

Table 11.6 Nutrient runoff and soil losses due to different mulching rates

Treatment	Available P (kg/ha)	Available K (kg/ha)	Total N (kg/ha)	Organic matter (ha-mm)	Runoff	Soil loss (t/ha)
Bare (T_0)	0.019	0.677	8.694	165.375	59.420	2.835
No mulch (T_1)	0.011	0.559	5.188	70.144	39.287	2.192
2 t/ha (T_2)	0.006	0.281	3.418	38.103	30.929	1.335
4 t/ha (T_3)	0.005	0.240	2.799	36.607	29.010	1.615
6 t/ha (T_4)	0.002	0.130	1.553	24.987	26.578	1.013
LSD (0.05)	0.0027	0.010	0.775	21.660	22.151	1.774

Source: Quansah *et al.* (2000)

Table 11.7 Maize grain yield under slash-and-burn and no-burn

Land-use stage	Field type	Grain yield (t/ha^{-1})
Annual cropping	*Monocrop maize*	
Slash, burn, no fertilizer	Monocrop maize	1.9
Slash, burn, with fertilizer	Monocrop maize	3.2
Slash, no-burn, no fertilizer	Monocrop maize	3.9
Slash, no-burn, with fertilizer	Monocrop maize	5.7

Source: GGDP (1992)

Moreover, nutrients applied to the soil in the form of mineral fertilizers are far less effective on soils in which organic matter has been lost than those which contain adequate amounts of it. The implication is that if the losses of N, P, and K were to be replenished by applying 15-15-15 compound fertilizer the desired effect on crop yield would hardly be attained because of low soil organic matter content. It has therefore been advocated that soil fertility replenishment in this region and in Africa in general should aim at integrated nutrient management (Quansah, 2000; Sanchez *et al.*, 1997). This involves the combined use of organic and inorganic inputs for sustaining soil fertility and crop yield.

Maize grain yield under slash-and-burn and no-burn management systems

The results shown in Table 11.7 (GGDP, 1992) amply demonstrate the yield advantage of no-burn over that of slash-and-burn. Maize grain yield under no-burning was 51 per cent higher than that under slash-and-burn. No-burning without mineral fertilizers was even better in grain yield than slash-and-burn plus fertilizers. It is further shown that applying mineral fertilizers could significantly enhance the higher grain yield under slash-and-no-burn.

Residual effect of slash-and-burn and no-burn on the yield of a maize/cassava intercrop

Table 11.8 shows the residual effect of slash-and-burn and mulching on the yield of a maize/cassava intercrop over a three-year period. The land-use study was carried out at Nkawie under the same land-use stage and field type as presented in Tables 11.4 and 11.5.

Table 11.8 Maize grain and cassava tuber yield under slash-and-burn and no-burn land management systems at Nkawie

Land-use stage	Field type	1996	1997	1998
Annual cropping	*Intercrop*		*Tonne per hectare*	
Slash-and-burn, no external input (T_0)	Maize	–	2.04	1.92
	Cassava	–	14.75	13.88
Slash, no-burn, external input (T_1)	Maize	2.18	2.02	1.94
	Cassava	10.74	13.50	15.00
Slash, no-burn with 4 t/ha^{-1} (T_2)	Maize	3.16	3.02	3.16
Poultry manure (PM) (approx. 100-50-64 kg NPK)	Cassava	22.13	19.54	24.94
Slash, no-burn with 4 t/ha^{-1} PM (T_3)	Maize	3.62	3.35	3.26
plus 30-20-20 kg NPK	Cassava	21.67	22.20	27.20
LSD (5%)	Maize	0.49	0.46	0.14
	Cassava	2.65	4.55	7.19

Source: Amoakohene (1999)

Over the three-year period, 1996–1998, there was a general decline in maize grain yield for all the treatments. Although the treatments were applied yearly, the added nutrients could not sustain the initial yield of the three-year continuous maize, which fed from the same rooting zone. On the other hand, because maize is a short-season crop, the slow release of nutrients from organic matter (poultry manure and vegetative biomass) may have failed to supply the nutrients needed to synchronize with the optimum demand of maize for nutrients. In such a situation, the yield of maize may decline.

Nevertheless the T_2 and T_3 treatments sustained a basal grain yield of 3 t/ha over the years. This is significantly higher than the average grain yield of 1 t/ha on most smallholder farms in Ghana. Although the period of yield monitoring was short, it is apparent that the T_2 and T_3 treatments hold a better promise than T_0 and T_1 for sustaining higher grain yield in a continuous cropping of maize.

In the case of cassava, there was a steady increase in tuber yield for T_1 and T_3 with the latter treatment recording the highest yields over the years. The observed steady increase in tuber yield over the years in respect of T_1, T_2, and T_3 might be due to the fact that the crops stayed on the field for a longer period, thus making optimum use of the nutrients mineralized and released from the organic matter.

Burning (T_0) gave the lowest tuber yield, which declined over the years as observed for maize grain yield. Burning could therefore not sustain higher crop yields.

Economic viability of the slash-and-burn and slash-and-no-burn management systems

A simplified financial analysis carried out by Boa-Amponsem (2002) shows the profitability of slash-and-burn and slash-and-no-burn management systems on farmers' field at Nkawie. The land-use type was annual cropping with a field type of monocrop maize. The mulching material was a mixture of *Chromolaena odorata* and *Panicum maximum*. The data were based on yields of the 2000 minor season. The partial budget indicated net returns in cedis of 2,010,000 (US$268) and 2,625,000 (US$350) for slash-and-burn and slash-and-no-burn respectively. The latter practice therefore increased net returns by 23 per cent.

State of the art in the practice of proka

Proka as practised today has undergone variable changes in its salient features. In its original form fire was only used to aid the felling of selected big trees during land preparation.

Currently, slash-and-burn is used to initiate the practice of *proka*. The underpinning factor for this shift is a response to the present high cost and scarcity of farm labour for land clearing, management of the large amounts of biomass, planting, and the subsequent maintenance of the fields. The introduction of light-demanding crops, particularly maize, as an intercrop in agroforest and orchard land-use stages is also a contributory factor. This is a food security strategy to increase grain stocks of the household and optimize the use of the land. Zero tillage introduced in areas where the vegetation is thick is often initiated by the slash-and-burn practice. The subsequent regrowth of herbaceous plants becomes more amenable to the practice of zero tillage.

Conclusions and recommendations

The study has demonstrated that *proka* as a land management practice conserves biodiversity and enhances soil biota. It enhances soil fertility through nutrient recycling and reduction of erosion and runoff. These attributes, together with enhanced soil moisture conservation and soil physical properties, contribute significantly to sustained soil productivity and crop yields.

In spite of these benefits, *proka* is increasingly dying off. There is therefore an urgent need to revive and encourage the use of the *proka* system in order to attain the goal of biodiversity conservation, sustainable soil and crop productivity, and environmental quality enhancement. Improvements in *proka* can be achieved by introducing soil amendments, especially manure and other sources of organic fertilizers, into the system. Further improvements can be achieved through the promotion of integrated nutrient management.

REFERENCES

Amoakohene, J. S., "The impact of different soil management systems on soil chemical properties and the yield of a maize/cassava intercrop: On-farm studies", BSc dissertation, Department of Crop Science, Kwame Nkrumah University of Science and Technology, Kumasi, 1999, unpublished.

Boa-Amponsem, K., "The effect of mulch management, seed treatment and planting techniques on stand establishment and maize grain yield under no-tillage", PhD thesis, Department of Crop Science, Kwame Nkrumah University of Science and Technology, Kumasi, 2002, unpublished.

GGDP, *Ghana Grain Development Project, 14th Annual Report*, Kumasi: Crops Research Institute, 1992.

Kang, B., and V. Balasubramanian, "Long term fertilizer trials on alfisols in West Africa", in *Transactions of XIV International Soil Science Society Congress*, Kyoto, Japan: ISSS Vol. 4, 1990.

Lal, R., *Soil Erosion problems on an Alfisol in Western Nigeria and Their Control*, IITA Monograph 1, Ibadan, 1976.

Quansah, C., "Integrated soil management for sustainable agriculture and food security in Ghana", in *FAO-RAF 2000/01, Integrated Soil Management for Sustainable Agriculture and Food Security: A Case Study from 4 Countries in West Africa (Burkina Faso, Ghana, Nigeria, Senegal)*, Accra: FAO Regional Office for Africa, 2000, pp. 33–75.

Quansah, C., E. Asare, E. Y. Safo, E. O. Amopontuah, N. Kyei-Baffour, and J. A. Bakang, "The effect of poultry manure and mineral fertilizer on a maize/cassava intercrop in peri-urban Kumasi, Ghana", in P. Drechsel and L. Gyiele, eds., *On-Farm Research on Sustainable Land Management in Sub-Saharan Africa: Approaches, Experiences and Lessons*, Bangkok: IBSRAM Proceedings, No. 19, 1998, pp. 73–90.

Quansah, C., E. Y. Safo, E. O. Ampontuah, and A. S. Amankwah, "Soil fertility erosion and the associated cost of N, P, and K removed under different soil and residue management in Ghana", *Journal of Agricultural Science*, Vol. 33, 2000.

Quansah, C., P. Drechsel, B. B. Yirenkyi, and S. Asante-Mensah, "Farmers' perception and management of soil organic matter – A case study from West Africa", *Natural Cycling in Agroecosystems*, Vol. 61, 2001, pp. 205–213.

Quansah, C., M. Bonsu, S. K. Agodzo, P. Gyawu, and S. Dittoh, *National Action Programme to Combat Desertification in Ghana*, Accra: Environmental Protection Agency, 2002.

Roose, E. J., *Natural Mulch or Chemical Condition for Reducing Soil Erosion in Humid Tropical Areas*, Special Publication, Soil Science Society of America Proceedings, No. 7, 1976, pp. 131–138.

Sanchez, P. A., R. J. Buresh, F. R. Kwesiga, A. U. Mokwunye, C. G. Ndritu, K. D. Shepherd, M. J. Soule, and P. L. Woomer, *Soil Fertility Replenishment in Africa: An Investment in Natural Resource Capital, Proceedings of the International Seminar on Approaches to Replenishing Soil Fertility in Africa – NGO perspectives*, Nairobi: ICRAF, 1997.

Zarin, D. J., G. Huijin, and L. Enu-Kwesi, "Methods for the assessment of plant diversity in complex agricultural landscapes: Guidelines for data collection and analysis from the PLEC Biodiversity Advisory Group (PLEC – BAG)", *PLEC News and Views*, No. 13, 1999, pp. 3–16.

12

Managing the home garden for food security and as a germplasm bank

Lewis Enu-Kwesi, Edwin A. Gyasi, and Vincent V. Vordzogbe

Introduction

As has been observed elsewhere, "A major global challenge, especially in developing countries, is to increase and secure food production for a growing population while, at the same time, conserving the natural diversity of crops, livestock, trees, and other life forms in their natural state" (Gyasi, 2002: 245). Promotion of home gardens on the basis of a sound understanding of their organization or structure and functioning is a possible way of meeting this challenge.

Simply defined, a home garden is any small parcel of land cultivated in the immediate neighbourhood of the residence of the cultivator or gardener. The process of cultivating such land is home gardening. It is done for either pleasure, or for economic reasons, or for both.

In addition to their small size and nearness to the home or point of consumption, home gardens are generally characterized by:

- the cultivation of a diversity of crops including vegetables and spices, food crops, tree crops, fruits, medicinal plants, and flowers and other ornamental plants
- a more or less permanent, continuous, or high-intensity cultivation that often involves multiple cropping, artificial watering of the plants, and high inputs of labour, manure, household refuse, and other forms of fertilizer.

They include "kitchen" gardens operated in the backyard for the primary purpose of producing vegetables and other crops that are frequently required by the kitchen or household of the operator, and other gardens located not too far away. Some of the plants, notably vegetables, are perishable by nature, others such as condiments are in frequent demand, and some, notably medicinal ones, are often needed at short notice for emergency purposes, hence the desirability of having the home garden near the home, the point of consumption. The locational pattern is in accord with von Thunen's model of land use (Chisholm, 1962).

The significance of home gardens lies in their role as a readily accessible source of food crops and other useful plants, and as a repository of biodiversity. Thus, properly nurtured home gardens stand to enhance both food security and germplasm.

Historically, home gardens have formed an important feature of the agricultural landscape in West Africa, in rural as well as urban areas.

Based on field studies carried out in southern Ghana under the United Nations University project on People, Land Management, and Environmental Change (UNU/PLEC), this chapter discusses home gardens with emphasis upon their implications for food security and conservation of plant genetic diversity, *in situ*, in agriculturally managed areas.

Methodology

Relevant information was obtained from field studies carried out in Gyamfiase-Adenya, Sekesua-Osonson, and Amanase-Whanabenya (Maps B, C, D, E) by the following methods:

- a systematic broad survey of a large number of home gardens
- a more in-depth investigation of a selected few
- impressionistic observations and personal interactions with farmers.

The survey was carried out on a multidisciplinary basis by a team of scientists (a botanist, soil scientist, and geographers having different specializations) with the support of technicians, graduate students, and local farmers. It involved the use of a questionnaire (Table 12.1) to determine floral, edaphic, yield, spatial, and socio-economic characteristics of gardens in the demonstration sites, each of which measures approximately 100 sq. km. In each case, the survey was preceded by explanation of the purpose through the chief, and by public announcement through the traditional village or town crier, *dawurubofo* (word in Akan-Twi).

In Gyamfiase-Adenya, the survey aimed at a comprehensive coverage of all the generally nucleated or agglomerated settlements and their living compounds houses (Table 12.2). This was achieved for about 50 per cent of the area.

However, in the case of Sekesua-Osonson, it was not considered completely feasible to aim at such a comprehensive coverage because of the difficult trekking problems posed by the elongated nature of the linear settlements founded on the *huza* landholding system. Therefore, after consultations with the leadership of the local association of PLEC farmers, the approach was modified. It involved

Table 12.1 Questionnaire for survey of home gardens

VILLAGE: Name........... DATE........... Compound No. .... Population.............

PLANT USE: Medicinal (M) Food (F); Fibre (B); Dye/food colour (D); Other...

A. Spatial and socio-economic situation	B. Name and use of plants including crops (name/list according to dominance)
1. Location of compound (a) Outskirts (b) Central 2. Home garden owner (a) Name........... (b) Gender.............. 3. Principal operational/caretaker (a) Name........... (b) Gender............. 4. Pacer's name: 5. Size of garden (no. of paces around perimeter)............................. 6. Borders of garden (a) Fenced (b) Hedged (c) Open	Food crops 1...................... 11................................. 2...................... 12................................. 3...................... 13................................. 4...................... 14................................. 5...................... 15................................. 6...................... 16................................. 7...................... 17................................. 8...................... 18................................. 9...................... 19................................. 10..................... 20.................................
7. Site and yield characteristics (tick) a) Refuse dump .... b) Soils Rich......... Moderate.......... Poor...... c) Yield High......... Moderate......... Low.......	Comments

progression by a time-saving systematic sampling procedure according to the three principal operational sectors of the demonstration site, namely:

- Bormase Whernya/Ternya, Bormase Dorse, and Siblinor in the eastern sector
- Sekesua and Dawa-Agbom in the central sector
- Prekumase, Osonson-Korlenya, Osonson-Yite, and Osonson-Sisi in the western sector (Map D).

This approach had the added merit of ensuring a reasonable spatial spread of the sample.

Subsequently, local farmer-assistants drawn mostly from the leadership of the PLEC farmers' association were assigned to determine the total numbers of settlements, of longitudinal landholdings, and of those holdings containing compound houses, to serve as a basis for determining appropriate sample size and interval for each of the sectors. After this, the survey proceeded along the principal road from one end of a settlement towards the other end, at an appropriate sample interval, according to a staggered or a straightforward systematic formation, depending upon whether or not compound houses were located at either side of the road.

Table 12.2 Distribution of home gardens and associated settlements and compound houses

Demonstration site and settlement covered	Population size (actually counted or estimated by sampling)	Total number of all compound houses/homes	Total number of *huza* longitudinal landholdings	Interval for systematic sampling	Number of compound houses sampled	Number of all home gardens; or of those in sampled compounds/homes
Gyamfiase Adenya						
Otwetiri	998	90	NA	NA	NA	42
Obom	282	41	NA	NA	NA	32
Adakaa	168	12	NA	NA	NA	111
Mampong-Nkwanta	376	45	NA	NA	NA	40
Kofi Noma	108	10	NA	NA	NA	4
Nyetia	61	6	NA	NA	NA	3
Akokoa	134	13	NA	NA	NA	11
Duasin	237	23	NA	NA	NA	19
Korkormu	566	67	NA	NA	NA	26
Sokoda-Guaso	153	15	NA	NA	NA	21
Addo-Nkwanta	518	84	NA	NA	NA	60
Yensiso	763	91	NA	NA	NA	64
Kwamoso Junction	172	20	NA	NA	NA	13
Gyamfiase	57	6	NA	NA	NA	5
Adenya	255	30	NA	NA	NA	22
Bewase	377	55	NA	NA	NA	35
Total	5,225	608	NA	NA	NA	408

Sekesua-Osonson						
1. Bormase Whernya/Ternya (E. sector)	915	88	122	9:12	10	8
2. Bormase Dorse (E. sector)	248	31	64	1:16	4	2
3. Siblinor (E. sector)	340	37	49	1:8	6	8
Subtotal	1,503	156	235		20	18
4. Dawa-Agbom (C. sector)	125	15	25	1:1	25	4
5. Sekesua (C. sector)	1,785	238	62	1:5	13	15
Subtotal	1,910	253	87		38	19
6. Prekumasi (W. sector)	590	50	74	1:12	6	6
7. Osonson-Korlenya (W. sector)	488	40	47	1:9	5	11
8. Osonson-Yiti (W. sector)	870	87	89	1:11	8	7
9. Osonson-Sisi (W. sector)	558	62	59	1:15	4	4
Subtotal	2,506	239	269		23	28
Total	3,971	648	591		81	65
Amanase-Whanabenya						
10. Whanabenya-Nyamebekyere	525	35	24	1:2	11	17
11. Aye-Kokooso	494	27	NA	NA	9	11
12. Obongo	322	28	21	1:1	20	23
13. Amanase	4,697	439	NA	NA	40	32
14. Aboabo	444	44	NA	NA	20	15
15. Abenabo	978	77	45	1:3	15	14
Total	7,460	650	90	–	115	112

NA: Not applicable

In Amanase-Whanabenya demonstration site, the settlement morphology is nucleated in Amanase and other areas owned by Akuapem migrant farmers, and is, as in the case of Sekesua-Osonson, linear in Whanabenya and other areas owned by migrant Siade/Shai farmers. Accordingly, in the selected Siade settlements (Whanabenya-Nyamebekyere, Obongo, and Abenabo; Table 12.2), the survey followed both systematic and purposeful sampling, depending upon layout of the houses. As in the case of Sekesua-Osonson, in Amanase-Whanabenya the choice of settlements was made in consultation with the leadership of the local PLEC farmers' association.

On the basis of the preceding methodology, in the survey as many as 804 compound houses distributed among 31 settlements came to be involved (Table 12.2), which provides a solid basis for making generalizations.

Each compound involved was visited and, using the recording sheet noted above (Table 12.1), the owner or operator of its home gardens interviewed on relevant aspects of the home garden management and the number of people resident in the compound. Home garden size was estimated by pacing around the perimeter. A rough-and-ready idea of plant species and of soil and topographical conditions was obtained by observation by eye. The auger was used to take samples of soil for laboratory analysis.

The survey probed further by selecting for closer study a few of the gardens surveyed around three of the compounds in the demonstration sites. Key methods involved were:

- plants diversity assessment by quadrats (Zarin, Huijin, and Enu-Kwesi, 1999)
- observations on the utility of the identified plants
- assessment of economic and remunerative potential of crops and other useful plants.

Following is a discussion of selected aspects of the huge amount of information generated from the fieldwork.

Management and organizational characteristics

On the whole, the gardens are small in size and located at the outskirts where there is more room.

Size ranges from less than 0.2 ha (0.5 acres) in a majority of cases to about 0.6 ha (1.5 acres). The smallness of size is in accord with the universal pattern. It is explained by competition with housing for land, by a need for intensive care, and for protection from goats and other foraging domestic livestock on the loose, particularly in a situation where gardens are generally not fenced off or effectively hedged (Table 12.3).

In Sekesua-Osonson and portions of Amanase-Whanabenya (especially Whanabenya, Nyamebekyere, and Obongo) that are occupied by Adangbe-speaking

Table 12.3 Management and organizational characteristics of home gardens in PLEC demonstration sites in southern Ghana (shown by percentage)

	Location		Ownership by gender		Gender of principal manager/ operator		Management according to landholding		Refuse dumped?		Fenced, hedged, open?				
Demonstration site	Out-skirts	Central	M	F	M	F	Owner managed	Tenant managed	Y	N	Fe	H	O	HO	FO
Gyamfiase-Adenya	78	22	78	22	80	20	51	49	71	29	14	6	78	1	1
Sekesua-Osonson	91	9	94	6	94	6	100	0	66	34	3	3	94	0	0
Amanase-Whanabenya	87	13	77	23	74	26	57	43	76	24	8	2	89	1	0
Average	85	15	83	17	83	17	69	31	71	29	8	4	87	1	0

M – Male; F – Female; Y – Yes; N – No; Fe – Fenced; H – Hedged; O – Open; HO – Both hedged and open; FO – Both fenced and open.

people on basis of the linear *huza* landholding arrangement, home gardens show significantly higher concentrations at the outskirts. This situation is associated with their linear housing pattern, which, typically, has no core zone of concentrated houses, unlike the pattern of nucleated settlements. The linear housing pattern appears to favour home gardening because from a homestead, a garden may be extended outwards to open lands uninhabited by other houses.

The ownership of home gardens predominantly by males (Table 12.3) accords with the traditional pattern whereby among Ghanaian societies farms, land, and other valuable property are generally owned or held in trust for the common good by the males.

However, the finding that the home gardens are managed or operated overwhelmingly by males should be interpreted with caution. Most probably this finding reflects a male-biased response to the questions, which were addressed mostly to the males because of their position as household or family heads. In the fieldwork, females were observed to be playing a central role in home garden management, especially by:

- weeding
- sustaining soil fertility by the common practice of dumping household refuse in the gardens or sweeping it into them
- tending vegetables and legumes raised for the household kitchen
- harvesting crops
- shooing away destructive livestock.

Although home gardens are mostly owner-managed, there is a significant representation of tenant-managed ones, except in Sekesua-Osonson where they are completely absent (Table 12.3). This absence is attributed to relatively limited number of "stranger" migrant-settler-tenant farmers in Sekesua-Osonson.

Diversity of crops

A large diversity of plants was encountered in the home gardens. As shown by Figure 12.1, in terms of regularity of species in the sample plots, the foremost ones are fruit plants, namely *Musa* spp. (plantain and banana) and *Citrus* spp. (orange). The second most important group comprises *Carica papaya* (pawpaw), *Elaeis guinensis* (oil-palm), *Ananas comosus* (pineapple), and economic woody trees such as *Funtumia africana* and *Newbouldia laevis*. The third order is made up of the food crops, *Xanthosoma maffafa* (cocoyam), *Manihot esculentum* (cassava), *Dioscorea* spp. (yam varieties), and economic woody species e.g. *Mangifera indica* (mango) and *Annona senegalense* (sweet apple), as well as weeds. The fourth and fifth ranks show relatively infrequent domesticated species.

From the preceding, the home gardens may, generally, be said to be of the agroforestry type that contain a diversity of plants dominated by edible fruits, food crops, and other useful plants.

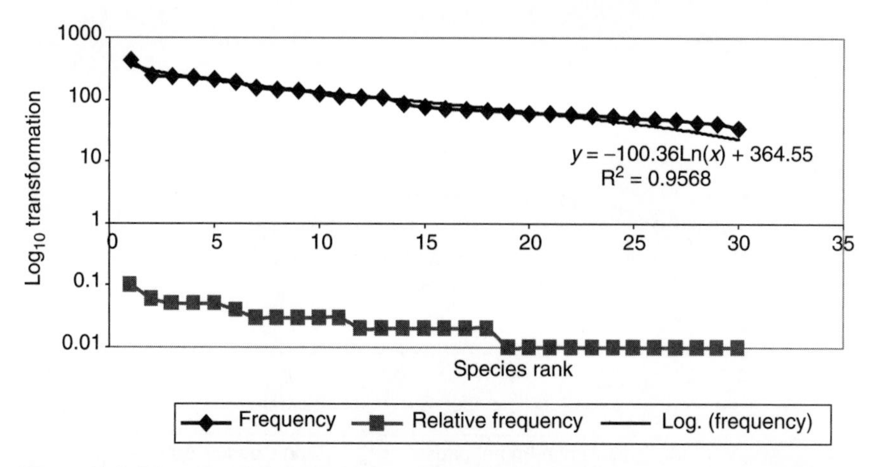

Figure 12.1 Diversity of plants in home gardens in 750 compounds in Gyamfiase-Adenya, Sekesua-Osonson, and Amanase-Whanabenya demonstration sites

The case of Odorkor Agbo's home garden

Among the home gardens selected for a closer study was the approximately 0.2 ha (0.5 acres) one managed on an owner-occupied basis by the PLEC farmers Odorkor Agbo and his wife Akwele at Adwenso in Sekesua-Osonson (Plate 5). The chapter focuses on it for deeper insights into home gardening as a system of securing both germplasm and food.

Species richness

A total of 83 different plant species belonging to 37 families were recorded. Four plant families, namely *Apocynaceae*, *Dioscoreaceae*, *Mimosaceae*, and *Sapindaceae*, dominated. On the average, a total of 27 species was recorded per 0.03 ha (0.074 acres) plot, i.e. 10 25 m^2 plots. The similarity of species among all 10 plots studied was 48 $\pm$ 1 per cent ($P > 0.05$ significance). Figure 12.2 shows the weighted averages of the different life-form categories found in Table 12.4, and the species richness.

Uses and other values

Out of the 83 species recorded, 47 (equivalent to 57 per cent) were distinguished for their "actual" uses and value (Table 12.5). About 30 per cent of the 47 species have locally known uses in native medicine as anti-venom, blood tonic, enema, treatment of boils, coughs, stomach troubles, and fever, and in

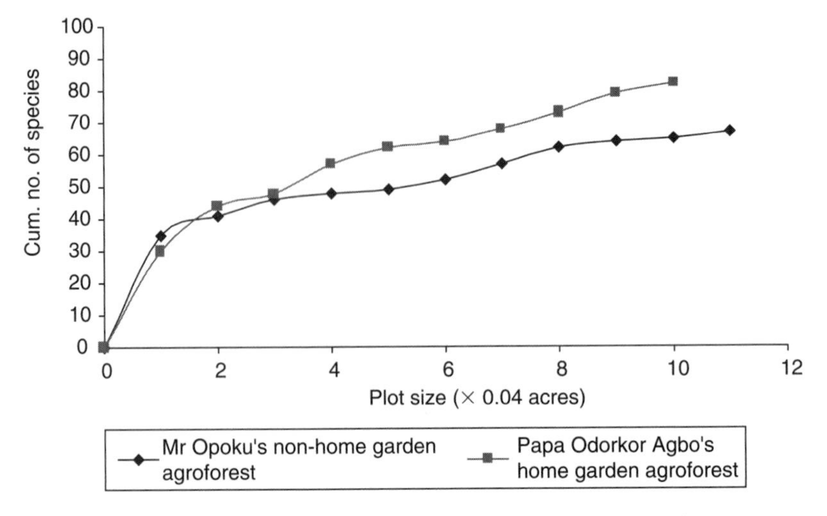

Figure 12.2 Species richness in Odorkor Agbo's home garden

the treatment of wounds and fresh cuts. The remaining 27 per cent constitute material for:

- craft making
- brooms for sweeping
- fans
- wooden ladles
- stool carvings
- pestles and mortar for pounding *fufu*, a local dish, and palm nuts
- twine for basket weaving
- rope for tying farm produce.

Other useful products include:

- timber for thatch-house construction
- fuelwood for domestic cooking and alcohol distillation

Table 12.4 Weighted averages of life-forms in Odorkor Agbo's home garden

Life-form	Types	Percentage importance
Erect shrubs	10	13 ± 4
Liane	6	10 ± 3
Trees and small trees	46	15 ± 3
Climbing herbs	1	4
Climbing shrubs	14	16 ± 3
Woody shrubs	1	4
Food crops	5	16 ± 5
Fruit trees	8	9 ± 3

Table 12.5 Use and other value of species in Odorkor Agbo's home garden

Plant species	Plant family	Normal life-form	Local name	Utility					Comment/remarks
				Medicinal	Craft	Food	Fuelwood	Other	
Acacia kamerounensis	*Mimosaceae*	Large liane	*Kotsa/dormelia*	–	–	–	–	Chewing sponge	A single large liane is worth ¢500 when processed
Adenia lobata	*Passifloraceae*	Large liane	–	–	–	–	–	–	–
Ageratum conyzoides	–	–	*Tokuwukolikpe*	Treatment of fresh wounds	–	–	–		Domestic use mainly
Albizia adianthifolia	*Mimosaceae*	Large tree	–	–	–	–	–	–	–
Albizia ferruginea	*Mimosaceae*	Large tree	*Pangpaa*	–	–	–	Fuelwood	Fodder	Good fuel-wood source
Albizia zygia	*Mimosaceae*	Medium-sized tree	–	–	–	–	–	–	–
Anona sp. (sweet type)	*Annonaceae*	Shrub/small tree	–	–	–	–	–	–	–
Anona sp. (sour type)	*Annonaceae*	Shrub/small tree	–	–	–	–	–	–	–
Antiaris toxicaria	*Moraceae*	Large tree	–	–	–	–	–	–	–
Azadirachta indica	*Meliaceae*	Medium-sized tree	–	–	–	–	–	–	–
Baphia nitida	*Papilonaceae*	Shrub/small tree	*Tungtso*	For treatment of coughs	Pestle	–	–	–	Medication is mainly domestic but pestle is sold
Blighia sapida	*Sapindaceae*	Medium-sized tree	–	–	–	–	–	–	–
Blighia welwitschii	*Sapindaceae*	Large tree	*Gbagblabata*	–	–	–	–	Yam-pole	Is good support

Table 12.5 (cont.)

Plant species	Plant family	Normal life-form	Local name	Utility					Comment/ remarks
				Medicinal	Craft	Food	Fuelwood	Other	
Bridelia ferruginea	*Euphorbiaceae*	Erect shrub	–	–	–	–	–	–	–
Calliandra sp.	*Mimosaceae*	Small climbing shrub	*Dedemotowe*	Enema	– –	– –	– –	– –	Domestic use mainly
Capsicum frutescens	*Solanaceae*	Woody shrub	*Kordayowi*	–	–	Spice	–		Domestic use mainly
Cardiospermum glandifolia	*Sapindaceae*	Small climbing herb	*Kpokpokpan*	–	–	–	–	–	–
Carica papaya	*Caricaceae*	Tree	*Gor*	–	–	Edible	–	Fodder for pigs	Domestic use mainly
Cassia sp.	*Caesalpinanceae*	Medium-sized tree	–	–	–	–	–	–	–
Ceiba pentandra	*Bombacaceae*	Large tree	*Leno*	–	–	–	–	Timber	Not sold
Centrosema rotunda	*Papilonaceae*	Small climbing shrub	–	–	–	–	–	Fodder	Domestic use mainly
Chassalia sp.	*Rubiaceae*	Small climbing shrub	*Niseiniieko*	For treatment of boils	–	–	–	–	Domestic use mainly
Chromolaena odorata	–	Erect shrub	*Acheampong*	–	–	–	–	Weed	Troublesome and widespread
Citrus spp.	*Rutaceae*	Small tree	*Kpebe*	–	–	Edible	–	–	Source of income
Clausena anisata	*Rutaceae*	Small tree	–	–	–	–	–	–	–
Clerodendrum sp.	*Verbenaceae*	Small tree	*Taagbesto*	–	–	–	–	Weed	–
Cnestis ferruginea	*Connaraceae*	Small climbing shrub	*Ngmleula*	Against “cowboil” and stomach pains	–	–	–	–	Domestic use mainly

Cocus nucifera	*Palmae*	Large tree	*Agorlegme*	–	–	Edible	–	Copra extracted for oil	Domestic use mainly
Cola millenii	*Sterculiaceae*	Large tree	*Asorkonabiatso*	–	–	–	–	Yam-pole	Is good support
Combretum sp.	*Combretaceae*	–	–	–	–	–	–	–	–
Cyathula prostata	*Amaranthaceae*	Erect shrub	–	–	–	–	–	–	–
Dialium guineensis	*Caesalpinaceae*	Medium-sized tree	–	–	–	Edible	–	–	Domestic use mainly
Dichapetalum madagascariense	*Dichapetalaceae*	Large liane	–	–	–	–	–	–	–
Dioscorea alata	*Dioscoreaceae*	Small climbing shrub	*Alamua*	–	–	Edible	–	–	Very tasty but delicate to maintain; 20 baskets per annum
Dioscorea sp. (wild type)	*Dioscoreaceae*	Small climbing shrub	*Odornor*	–	–	Edible	–	–	Very tasty and has market; 96–100 baskets per annum
Dioscorea sp. (wild type)	*Dioscoreaceae*	Small climbing shrub	*Cat*	–	–	Edible	–	–	Very tasty and has market
Dioscorea sp. (lighter yellow)	*Dioscoreaceae*	Small climbing shrub	*Kani*	–	–	Edible	–	–	Very tasty and has market
Dioscorea sp. (wild type)	*Dioscoreaceae*	Small climbing shrub	*Baale*	–	–	Edible	–	–	Very tasty and has market
Elaeis guinensis	*Palmae*	Tree	*Tang*	–	Broom, fan	Oil, alcoholic beverage	–	–	Domestic use mainly: palm-wine is tapped mainly to entertain visitors and friends
Ehretria cymosa	*Boraginanceae*	Tree	–	–	–	–	–	–	–

Table 12.5 (cont.)

Plant species	Plant family	Normal life-form	Local name	Utility					Comment/ remarks
				Medicinal	Craft	Food	Fuelwood	Other	
Euphorbia heterophylla	*Euphorbiaceae*	Erect shrub	–	–	–	–		Fodder	–
Ficus exasperata	*Moraceae*	Large tree	*Sabatso*	Treatment of thorn wounds	–	–	Fuelwood	–	Charcoal burns well
Fantasia africana	*Apocynaceae*	Large tree	–	–	–	–	–	–	–
Funtumia elastica	*Apocynaceae*	Large tree	–	–	–	–	–	–	–
Gloriosa superba	*Liliaceae*	Small climbing shrub	–	Delayed walking in infants	–	–	–	–	Medication is mainly domestic
Grewia carpinofolia	*Tiliaceae*	Large liane	–	–	–	–	–	–	–
Griffornia simplicifolia	*Caesalpinaceae*	Climbing shrub	Totolimo	–	–	–	–	Chewing stick	Seeds are usually harvested and sold
Hibiscus esculenta	*Malvaceae*	Erect shrub	–	–	–	–	–	–	–
Holarrhena floribunda	*Apocynaceae*	Tree	*Kiakin/oflutu*	–	Ladle and stools	–	Fuelwood	Yam-pole	Hard wood mainly for carving the chief's stool and gunboat
Hippocratea sp.	*Celastraceae*	Large liane	–	–	–	–	–	–	–

1. Biodiverse agroforestry in Gyamfiase-Adenya

2. *Proka*, a no-burn farming practice that involves mulching by leaving slashed vegetation to decompose *in situ*

3. Wooden beehive in a conserved forest

4. Emmanuel Nartey, an expert farmer standing in front of a wooden beehive in his agro-forestry home garden at Bormase, Sekesua-Osonson demonstration site

5. Odorkor Agbo, an expert farmer demonstrating harvesting of yam in his home garden managed by agroforestry principles at Adwenso, Sekesua-Osonson demonstration site

6. Cecilia Osei (middle), an expert in the *proka* mulching, no-burn system in a farm with other farmers and a PLEC scientist at Jachie demonstration site

7. Managing *Cassia siemens* (a popular wood for fire and charcoal) by coppicing

8. Bush yam, *D. praehensilis* (*obobi*)

9. A local tomato land-race cultivated within the compound house of a PLEC expert farmer, Emmauel Nartey (wife – standing to right, and Prof. Gyasi – left) at Bormase, Sekesua-Osonson demonstration site

10. A beehive made of an earthen pot kept within a forest conserved in the backyard, a traditional way of beekeeping

Justicia sp.	*Acanthaceae*	Erect shrub	*Mu*	Stomach pain, enema	–	–	–	–	–
Kigelea africana	*Bignoniaceae*	Large tree	*Mormotso*	–	–	–	Fuelwood	Yam-pole	Is good support
Lantana camara	*Verbenaceae*	Erect shrub	*Akotongme*	–	–	–	–	–	–
Lecaniodiscus cupaniodes	*Sapindaceae*	Tree	–	–	–	–	–	–	–
Mallotus oppositifolius	*Euphorbiaceae*	Tree	*Sofietso/sesetso*	Anti-venom	–	–	–	–	–
Mangifera indica	*Anacardiaceae*	Medium-sized tree	–	–	–	–	–	–	–
Manihot esculentum	–	–	*Agbeli*	–	–	Edible	–	–	A staple food crop
Millettia zechiana	*Papilionaceae*	Medium-sized tree	*Huntso*	–	–	–	–	Yam-pole	Is good support
Mirabilis jirapa	–	–	–	–	–	–	–	–	–
Mondia lucida	*Asclepiadaceae*	Small climbing shrub	–	Blood tonic	–	–	–	–	–
Monodora tenuifolia	*Annonaceae*	Medium-sized tree	–	–	–	–	–	–	–
Morinda lucida	*Rubiaceae*	Medium-sized tree	*Duortso*	–	–	–	–	–	–
Momordica charantia	*Cucurbitaceae*	Small climbing shrub	*Nyanyla*	Treatment of fever	–	–	–	Yam-pole	Is good support
Morus mesozygia	*Moraceae*	Tree	–	–	–	–	–	–	–
Musa spp.	*Musaceae*	Tree	*Manna*	–	–	Edible	–	–	–

Table 12.5 (cont.)

Plant species	Plant family	Normal life-form	Local name	Utility					Comment/ remarks
				Medicinal	Craft	Food	Fuelwood	Other	
Nesogordonia papaverifera	*Sterculiaceae*	Large tree	*Bano*	–	Gunboat, pestle, and mortar		Fuelwood	Timber	–
Newbouldia laevis	*Bignoniaceae*	Tall tree	*Nyabatso*	–	–	–	Fuelwood	Yam-pole	Is good support
Ocimum gratisimum	–	–	–	–	–	–	–	–	–
Paullinia pinnata	*Sapindaceae*	Small climbing shrub	*Aklokinga*	Treatment of dysentry	–	–	–	–	–
Pavetta sp.	*Rubiaceae*	Small climbing shrub	–	–	–	–	–	–	–
Persia americana	*Lauraceae*	Medium-sized tree	*Peya*	–	–	Edible	–	–	Has good market value
Rauvolfia vormitoria	*Apocynaceae*	Erect shrub	–	–	–	–	–	–	–
Rothmannia longiflora	*Rubiaceae*	Tree	–	–	–	–	–	–	–
Salacia sp.	*Celastraceae*	Large liane	*Aklade*	–	Rope	–	–	–	–
Sida acuta	*Malvaceae*	Erect shrub	*Torgetorge*	Treatment of boils	–	–	–	–	Domestic use mainly
Solanum torvum	*Solanaceae*	Erect shrub	*Gatsoku*	–	–	Edible	–	–	For house-hold cooking mainly
Spondias mombin	*Anacardiaceae*	Medium-sized tree	*Akorletso*	–	–	–	–	–	–
Sterculia trigacantha	*Sterculiaceae*	Large tree	*Torgordzor*	–	–	–	–	–	–

Theobrona cacao	*Sterculiaceae*	Medium-sized tree	–	–	–	–	–	–	–
Trema sp.	*Ulmaceae*	Tree	–	–	–	–	–	–	–
Trichilia sp.	*Meliaceae*	Tree	–	–	–	–	–	–	–
Trilepisium madagascariensis	*Moraeceae*	Tree	–	–	–	–	–	–	–
Vernonia conferta	*Asteraceae*	Tree	*Agba*	Treatment of metal wounds	–	–	–	–	Domestic use mainly
Voacanga africana	*Apocynaceae*	Medium-sized tree	–	–	–	–	–	–	Occasionally harvested and sold to outsiders. No known economic value
Xanthosoma mafaffa	*Araceae*	–	*Mankani*	–	–	Edible	–	–	–
Indet.	–	–	*Korkorteinye*	–	–	–	–	Bathing sponge	–

- bathing sponges
- chewing sticks
- fodder for backyard animals.

Young stalks of *Blighia welwetschii*, *Cola millenii*, *Horralhaena floribunda*, *Kigelia africana*, *Milletia zechiana*, *Morinda lucida*, and *Newbouldia laevis* are deliberately left standing on-farm to regenerate and to serve various purposes, including usage as live stakes for yams. Most of these trees have a root morphology that favours their use as traditional yam variety hold-up poles. They are usually deep-rooted, and do not form profuse lateral roots that can hamper growth and tuberization of yams. After attaining a certain level of maturity, they are harvested as fuelwood and timber.

It is estimated that a minimum amount of ¢1 million (= >US$222 at the prevalent $1 = ¢4,500 exchange rate) could be realized seasonally or annually from sales of items from Odorkor Agbo's garden as listed in Table 12.6. The potential marketing return of $222, which excludes the proportion of garden produce consumed directly by the Agbos, compares to Ghana's estimated $370 GDP per capita (African Development Bank, 1995). This finding suggests that if properly nurtured, home gardens would not only serve as a major reservoir of germplasm, but also form a significant basis of improving rural livelihoods and food security.

Other findings on utility

Non-food tree species are also found in home gardens. They include *Newbouldia laevis*, used as life stakes for yams, and *Funtumia africana*, used for carvings and artifacts for sale in addition to its known medicinal uses. Therefore, the actual and potential role of home gardens in securing food supplies, and in conserving plant genetic basis of the food production and marketing systems needs to be accorded greater policy attention.

Table 12.6 Major items that could be sold from Odorkor Agbo's garden

Product	Source
Pestles	*Horralhaena floribunda*
Chewings and bathing sponge sticks	*Acacia kamerunensis*
Yam tuber/seedlings	*Dioscorea* spp.
	• *Odornor* only
	• *Alamaa* only
	• Seedlings only
Cassava	*Manihot esculenta*
Fuelwood/charcoal	*Horralhaena floribunda*
	Leucaena sp./*Cassia* sp.
Leafy vegetables	*Xanthosoma maffafa* only
Fruits	*Persia americana* only
	Citrus sp. only

The second most distinguished home garden agroforest tree species identified is *Elaeis guinensis* (oil-palm). When fully developed and its fruits harvested, *E. guinensis* produces quality red oil for cooking. The stem sap is also tapped as a local beverage and often fermented and distilled into alcohol, a local gin popularly called *akpeteshie*. The leaf petiole and young fronds of *E. guinensis* are used for producing baskets and brooms.

Apart from provision of food, most other herbaceous and grass species have medicinal value. Classical examples encountered are non-woody herbs (not including grasses) in home gardens of PLEC farmers at Bewase and Otwitiri in Gyamfiase-Adenya, a second PLEC demonstration site in southern Ghana. In those localities the following species and their medicinal uses were identified:

- *Palisota hirsute*, whose leaves are used as purgative to remedy constipation in babies whilst the roots are used for treating bone fractures in both children and adults
- *Datura metel*, which is used as medicine for poultry
- *Chassalia* sp., whose roots are used for treating epilepsy
- *Canna indica*, used to control post-partum bleeding
- *Cyathula prostrata*, for treating stomach ache
- *Strophanthus hispidus*, whose marsh roots are used as an ear irritant to torture a criminal
- *Leea* sp. and *Scropar*ia sp., whose root bark is used for treating boils
- *Thonningia sanguina*, a plant-root parasite, whose stem is used as enema
- *Hilaria latifolia*, whose leaves are customarily used in fetish baths.

These observations on medicinal plants are significant because the use of herbal remedies is increasing amongst both educated and non-formally educated people in urban as well as rural areas, despite the availability of modern medicinal facilities such as hospitals, clinics, and health centres. An estimated 75 per cent of the country's population use herbal remedies, many of them based on home garden plants (UNDP, 1999).

Furthermore, there have been several recent expeditions in search of plant species with potential medicinal properties for treating currently incurable diseases, for example cancers and HIV/AIDS, or for use as templates or analogues to develop novel drugs that may be efficacious for these diseases. Home gardens are an important habitat for these potential medicinal plants. It is in recognition of this that the GEF Small Grants window provides support for the establishment of biodiverse medicinal plant gardens in the country.

Summary and conclusion

On the basis of the PLEC studies, the principal plant species (including tree and pseudo-tree crops) kept in home gardens are the *Musa* spp. (plantains and bananas), *Citrus* sp. (orange), *Carica papaya* (pawpaw), and *Elaeis guinensis* (oil-palm).

The reason is that they constitute an important carbohydrate and fat staple alongside cassava and yams.

Fleshy leaves of a few herbaceous species, for example, *Xanthosoma maffafa* (*kontomere* – local vernacular name), *Fleuya aestuanes*, *Talinum triangulare* (*bokoborkor*), and *Corchorus olitorus*, are kept in home gardens and have value as vegetables for cooking.

REFERENCES

African Development Bank, *Selected Statistics of Regional Member Countries*, Abidjan: ADB, 1995.

Chisholm, M., *Rural Settlements and Land Use: An Essay in Location*, London: Hutchinson University Library, 1962.

Gyasi, E. A., "Traditional forms of conserving biodiversity within agriculture: Their changing character in Ghana", in H. Brookfield, C. Padoch, H. Parsons, and M. Stocking, eds, *Cultivating Biodiversity: Understanding, Analysing and Using Agricultural Diversity*, London: ITDG Publishing, 2002, pp. 245–255.

UNDP (United Nations Development Programme) *Ghana Human Development Report (HDR) for 1999: Sustaining the Environment for Poverty Reduction: Biodiversity and Resource Utilization*, 1999, Accra, unpublished, pp. 33–54.

Zarin, D. J., G. Huijun, and L. Enu-Kwesi, "Methods for the assessment of plant species diversity in complex agricultural landscapes; Guidelines for data collection and analysis from the PLEC Biodiversity Advisory Group (PLEC-BAG)", *PLEC News and Views*, No. 13, 1999, pp. 3–16.

13

Management of trees in association with crops in traditional agroforestry systems

John A. Poku

Introduction

Generally agroforestry refers to land-use systems in which trees or shrubs are grown in association with agricultural crops, pastures, or livestock, and in which there are both ecological and economic interactions between the trees and other components (Young, 1989).

Historically, in most parts of West Africa and Ghana in particular, it is known that farmers and other land users practise agroforestry in one form or another (MOFA, 1989).

Shifting cultivation, bush fallowing, and trees on rangeland and pastures are some of the agroforestry practices that have often been referred to as traditional agroforestry systems. The trees in traditional agroforestry practices attracted little or no resources by way of management compared to what obtains in practices like alley cropping, fodder banks, and live fences. In recent times, however, farmers have developed minimum management practices aimed at enhancing the service or product roles such trees can offer.

Traditional agroforestry systems are found in all the major agro-ecological zones in Ghana. They form part of the traditional cropping systems.

This chapter does the following:

- describes the different traditional cropping and agroforestry systems in the various agro-ecological zones
- lists the tree species that are left *in situ* on farms

- states the rationale for and attributes of the above-mentioned practices
- highlights the relevant management practices.

In Ghana the five main agro-ecological zones are:

- forest
- forest-savanna transition
- Guinea savanna (southern sector of the interior savanna)
- Sudan savanna (northern sector of interior savanna)
- coastal savanna (Map B).

Forest zone

The forest zone covers the Western, greater parts of Ashanti, parts of Brong Ahafo, Eastern, and Volta regions and some parts of the Central region. These are administrative regions. As such, they are capitalized. It makes up roughly 33 per cent of the country's land mass, and comprises subclasses such as the wet evergreen, moist semi-deciduous, and dry semi-deciduous forests. It enjoys a bimodal rainfall pattern with a mean annual range of between 1,500 and 2,000 mm. Soils are dominated by forest ochrosols and oxysols.

Traditional cropping systems

Traditional cropping systems practised in the forest zone include mixed cropping, monocropping, and intercropping.

The mixed cropping system involves the growing of different crops on the same piece of land at the same time. Crop combinations found in the system include maize/cassava/vegetable and plantain/cocoyam/yam. These crop combinations can be found among the major cash crops such as cocoa, coffee, oil-palm, and citrus in the early stages of their cultivation. The mixed cropping system promotes crop diversity, enhances food security, and makes efficient use of land and labour.

In the case of monocropping, also known as sole cropping, a single crop is grown on a piece of land over one or two cropping seasons. Maize, rice, cassava, and tomato are usually grown in this system. The practice is however not widespread because of the economic and ecological risks associated with it.

Traditional agroforestry systems

In the forest zone, two main traditional agroforestry practices can be recognized, namely bush fallowing, and tree crop/cash crops/forest complex.

The bush fallow system basically involves the growing of food crops for a period extending between one and three years on newly cleared land, and leaving the land to fallow for between four and 10 years. During the fallow period, trees and shrubs that regenerate are expected to improve the soil fertility status of the land through nutrient recycling. At the end of the fallow period farmers return to the land and recultivate it with food crops. Food crops cultivated under this system depend on the farmers' choice, and climatic and soil conditions. Typical food crops grown are plantain, banana, cassava, yam, maize, and vegetables.

The food crops/tree crops forest complex system of traditional agroforestry is found in situations where food crops are cultivated under cash crops, which grow under naturally dispersed forest trees. The system forms a three-tier storey with the food crop at the lower level, the cash crop in the middle, and the forest trees in the upper storey. The cocoa-plantain/banana-*odum*/*wawa* complex typifies this system.

Rationale for leaving tree species in situ *on farms*

It has been observed that farmers practising the traditional cropping systems leave quite a number of tree species on the farmland, either intentionally or for convenience. Table 13.1 lists the most important tree species left *in situ* on farms.

Farmers leave trees *in situ* on farms for various reasons. The main reason is to conserve nature to derive future benefits and for immediate use, which includes food, fodder, fuelwood, medicine, construction, and shade, and many

Table 13.1 Tree species left *in situ* in farms in the forest zone

Scientific name	Common name
Chlorophora excelsa	*Odum*[1]
Terminalia superba	*Ofram*[1]
Triplochiton scleroxylon	*Wawa*[1]
Militia excelsa	*Odum*[1] (African teak)
Ceiba pentandra	*Onyina*[1] (silk cotton tree)
Terminalia ivorensis	*Emire*[1] (shingle tree)
Khaya ivorensis	Mahogany
Piptademistrum aficanum	*Dahoma*[1]
Nuclea diderrichii	*Kusea*[1]
Entandophragma angolense	*Sapele*[1]
Microdermis puberia	NK[2]
Albizia zygia	*Okoro*[1]
Ficus anomani	Ficus
Cola gigantea	*Owataka*[1] (wild cola)
Lophira lanceolata	NK[2]
Afzelia africana	NK[2]

1. Local/vernacular names.
2. Not known.

more. The practice of leaving trees *in situ* on farms ensures that basic farm resources are continually renewed, easily available, and, above all, provide diversity to the agricultural environment.

Farmer management practices

Generally croplands in the forest zone are routinely maintained through the various husbandry practices of weeding, pest control, and timely harvesting. In the special case of trees left *in situ* on farms, tree saplings are protected against weeds till they are strong enough to compete with weeds and other shrubs. There is no conscious effort to fertilize the trees or shrubs, but invariably they benefit from the minimal amounts of fertilizers that are applied in maize and vegetable fields. To reduce competition between the food or cash crop and the tree crop, regular tree side pruning, lopping, and pollarding are undertaken during cropping seasons.

Forest-savanna transition zone

The transitional zone is an intergrade between the forest and the savanna zones. It covers part of the Central, Eastern, Volta, and Ashanti regions and almost the entire Brong Ahafo region.

Traditional cropping systems

Basically the traditional cropping systems practised in the transitional zone are not different from those described for the forest zone. However, it is important to note that continuous cropping is widespread in the transition zone.

Maize-based farming is typical of this zone. There also are subsystems such as maize/rice/yam, maize/yam, and plantain/cocoyam intercrops, as well as vegetables. Other crops of prominence cultivated in the zone are tobacco, cowpea, and cassava. The cropping sequence and patterns take several forms. They include sole cropping, rotations, and multiple cropping.

Traditional agroforestry systems

The following traditional agroforestry systems are found in the zone:

- shifting cultivation/bush fallowing
- tree crop/food crop/forest complex
- plantation crop/root and tuber crop complex.

Compared to the bush fallowing in the forest zone, the transitional zone has shorter fallow periods.

Rationale for leaving tree species in situ *on farms*

Table 13.2 provides the names of the most common trees left *in situ* on farms.

The rationale for leaving trees *in situ* on farms is the same as in the forest zone. However, in the transition zone, on-farm trees play more crucial roles than in the forest zone. Trees on farmlands serve as shade for livestock. Mangoes in the zone are a key source of income for the farmers. Some of the trees also serve as wind-breaks for crops and farm settlements.

Farmer management practices

Trees on farms are mostly pruned to reduce shade on crops. Fruit trees such as cashew and mango are planted using recommended spacing. When such trees are inter-cropped with food crops such as maize, cowpea, and soya bean, they benefit from the husbandry practices associated with the food crops. In isolated cases, fallow lands with economic trees are protected against bushfire by fire belts and early burning.

Guinea savanna zone

The Guinea savanna zone covers the Northern region and parts of the Upper East, Upper West, and Volta regions. It forms the major part of the interior savanna zone of Ghana.

Table 13.2 Tree species left in situ in farms in the transitional zone

Scientific name	Common name
Ceiba pentandra	Silk cotton tree
Senna siamea	Cassia
Anogeissus leicocarpas	Anogeissus
Azadirachta indica	*Neem*
Blighia sapida	Akee apple
Anarcadium occidental	Cashew
Ficus exasperata	Ficus
Ficus spp.	Fig tree
Mangifera indica	Mango
Parkia biglobossa	*Dawadawa*[1]–African locust bean
Butyrospermum parkii (*Vitellaria paradoxa*)	Sheanut
Daniella oliveri	NK[2]
Chlorophora excelsa	*Odum*[1]

1. Local/vernacular name.
2. Not known.

Traditional cropping systems

Cropping systems identified in the zone are crop rotation, mixed cropping, sole cropping, and intercropping.

An important feature of the cropping systems in the Guinea savanna zone is the prevalence of compound farms and bush farms. The compound farm type operates around human settlements and typifies communities in which cropland is scarce. Bush farms are rather distant farms normally located two to three kilometres away from the homestead. This practice operates in areas where land is relatively abundant.

The zone features the maize/cattle and sorghum-based systems, with diverse subsystems that involve crops such as yam, maize, sorghum, millet, rice, cowpea, groundnut, and vegetables in different combinations.

Traditional agroforestry systems

The Guinea savanna has a distinct traditional agroforestry system that integrates annual crops, tree crops, and livestock. The system can be delineated into two main practices. These are dispersed trees on croplands and dispersed trees on rangelands. Minor practices identified in the zone include boundary planting, live fencing, and home gardens.

Rationale for leaving trees in situ *on farms*

Table 13.3 lists the names of tree species that are left *in situ* on farms in this zone.

Most of them provide fruits and nuts that are either sold or consumed in the household. This is the case especially with the *dawadawa* and shea butter trees. Additionally, farmers use some of the trees for construction of homesteads, farm implements, and local bridges. Medicinal as well as spiritual values are also associated with some of the trees.

Management practices

The *dawadawa*, shea butter, and albizia trees are often associated with crops. It is alleged that some farmers influence the spatial distribution of such trees in order to reduce the shading effects of the trees on the associated crops. Uprooting young saplings of such trees to achieve not only a desired tree population but also an acceptable spatial arrangement does this. In the case of mature trees on croplands, lopping, removal of dead branches, and pruning are some of the tree management practices.

Table 13.3 Tree species left *in situ* in farms in the Guinea savanna zone

Scientific name	Common name
Parkia biglobosa	West African locust bean
Adansonia digitata	Baobab
Ceiba pentandra	Kapok
Butyrospernum parkii (V. paradoxa)	Shea butter
Balanites aegyptica	–
Vitex doniana	Blackberry
Faidherbja albida	–
Afzelia africana	–
Diospynis mespiliformis	Ebony tree
Gerdenia spp.	–
Ficus spp.	–
Tamarindus indica	Tamarind
Lannea acida	–
Isoberline spp.	–

Sudan savanna zone

The Sudan savanna zone covers the north-eastern corner of the Upper East region and the northern fringes of the Upper West region.

Traditional cropping systems

In general, agricultural land in the Sudan savanna zone is poor in quality and limited in quantity. In the Upper East portion of the zone, compound farming with continuous multiple cropping is common. The zone also forms part of the millet-based system in which major crops such as millet, sorghum, groundnut, and cowpea are cultivated. In addition, some minor crops such as soya beans, maize, and a variety of vegetables are cultivated. In the Upper West portion of the zone, cassava cultivation in compound gardens is very prominent.

Traditional agroforestry systems

The most prominent practice in terms of traditional agroforestry is dispersed trees on croplands. The most common tree species include *neem*, shea butter, and *Faidherbia albida*.

Rationale for leaving trees in situ *on farms*

Table 13.4 lists tree species left *in situ* in farms. It is significant to note that the tree species do not differ from those identified under the Guinea savanna zone.

Table 13.4 Tree species left *in situ* in farms in the Sudan savanna zone

Scientific name	Common name
Parkia biglobosa	West African locust bean
Adansonia digitata	Baobab
Ceiba pentandra	Kapok
Butyrospernum parkii	Shea butter
Balanites aegyptica	–
Vitex doniana	Blackberry
Faidherbia albida	–
Afzelia africana	–
Diospynis mespiliformis	Ebony tree
Gerdenia spp.	–
Ficus spp.	–
Tamarindus indica	Tamarind
Lannea acida	–
Isoberline spp.	–

The major difference in these two zones is the density of trees, which is higher in the Guinea savanna zone.

Trees are left *in situ* on farm for various reasons. The most important reasons are to:

- secure edible leaves, fruits, and seeds for human as well as livestock consumption
- provide shade for humans and livestock
- provide material for medicinal purposes
- serve as landmarks for farm and property boundaries
- control soil erosion and improve soil fertility.

An interesting attribute of the practice is that some of the tree species are self-propagating and, thus, human labour for planting is minimized. Another is that the evergreen nature of the species ensures a constant supply of fodder and browsing material for livestock. The phenology of the tree species, e.g. *Faidherbia albida*, allows crop cultivation at the appropriate time. The tree management practices are similar to those in the Guinea savanna zone.

Coastal savanna zone

The coastal savanna is located mainly in the coastal strip between Sekondi and the Ghana-Togo border. It covers the Greater Accra region, and parts of the Volta, Central, Eastern, and Western regions.

Traditional cropping systems

The cropping system is typically cassava based. Within this broad system are the subsystems of cassava/livestock and vegetable/cassava. Patterns of crop cultivation are mixed or monocrop depending on the farmers' circumstances and weather conditions. Because of land shortage, continuous cropping is widely practised.

Traditional agroforestry systems

The traditional agroforestry systems in the coastal savanna zone assume the form of trees scattered on rangelands and those on croplands. The first one, which involves trees scattered on rangelands, is prevalent in the Accra plains. Trees on the range serve as fodder and shade for livestock. *Griffonia simplicifolia* is noted for its fodder, and medicinal seeds for the export market.

The second, which involves trees on cropland, also features as a traditional agroforestry practice in the zone. Additionally, mangrove management is an age-old practice along some of the coastline. In the Keta district in particular, the inhabitants harvest the mangrove for fuelwood for fish smoking. There is minimal planting of mangrove to replenish the stock.

Rationale for leaving trees in situ *on farms*

Farmers plant or maintain trees on their cropland primarily to obtain valuable tree products such as fuelwood, charcoal, fodder, food, medicinal products, and fibre (Table 13.5). Some also do so to restore fertility of the soil. The practice ensures sustainability of the particular resource base, enhances biodiversity in the agricultural environment, and makes the people self-reliant.

Management practices

Farmers generally carry out weeding on the croplands, which benefits the trees as well. Tree pruning, coppicing, and pollarding are also undertaken to reduce

Table 13.5 Tree species left *in situ* in farms in the coastal savanna zone

Scientific name	Common name
Adansonia digitata	Baobab
Azadirachta indica	*Neem*
Anogeissus leiocarpus	Anogeissus
Blighia sapida	Akee apple
Borassus africana	Borasus palm
Ceiba pentandra	Silk cotton
Dalium guinenses	Dalium
Fumtumia elastica	–
Griffonia simplicifolia	–
Mangifera indica	Mango
Spondias mombin	–

competition between crops and trees, and to achieve the desired form of the tree, e.g. rafter poles, timber, etc. Bushfire prevention measures are at times put in place. Various protective cages, ranging from wooden to metal, are employed to discourage livestock from trampling on young saplings. In some cases, livestock faeces and urine are sprinkled on trees to ward off browsing livestock such as goats and cattle.

Conclusion

Ghana abounds in traditional agroforestry systems that show great biodiversity. A challenge is to develop them in ways that ensure food security while conserving the biodiversity.

REFERENCES

MOFA (Ministry of Food and Agriculture), *Support for National Agroforestry Programme in Ghana: Consultancy in Agroforestry*, Accra: MOFA, 1989.

Young, A., *Agroforestry for Soil Conservation*, Oxford: CAB International, 1989.

14

Preliminary observations on effects of traditional farming practices on growth and yield of crops

Leonard Asafo, Ebenezer Laing, Lewis Enu-Kwesi, and Vincent V. Vordzogbe

Introduction

Scarcity of land, particularly for farming, has become a major problem in Ghana. It is attributed to increasing population pressure, which among other things results in deforestation and loss of biodiversity. Closely associated with the forest destruction are unsustainable farming practices such as the slash-and-burn land preparation methods and continuous cropping without adequate soil fertility regeneration measures.

Even so, there are traditional practices which, if encouraged, could help alleviate the problem. Those practices are founded on what has come to be variously known as traditional knowledge, local knowledge, and indigenous knowledge. Traditional knowledge refers to the knowledge, innovations, and practices of indigenous local communities. It is developed primarily from experience and is transferred from one generation to the next by oral tradition or word of mouth.

Agroforestry is an outstanding example of traditional practices based on an intimate understanding of local conditions. Since times immemorial, farmers have practised traditional agroforestry by growing crops in association with trees deliberately left unfelled.

In simple terms, agroforestry refers to the deliberate association of trees and other woody plants with crops or animals or both on the same piece of land. It has been suggested as a practice that can halt or drastically reduce the rate of deforestation, especially in the tropics.

As discussed in Chapters 6 and 11, a traditional farming system founded on agroforestry principles in Ghana is *proka*. *Proka* involves no burning after slashing, and makes use of the chopped vegetation for mulching. It continues to be practised widely, especially in the humid and semi-humid forest zones, even though its popularity appears to be on the decline.

The declining popularity of *proka*, despite its apparent ecological advantages and the fact that the ecology and economics of the system still remain inadequately understood, underscores a need for further research into it and other forms of traditional agroforestry. Such research holds a key to sustainable food production while maintaining natural biodiversity.

Study objectives

Accordingly, a project was designed to:

- carry out a preliminary survey of trees left on farmlands in selected areas within Gyamfiase-Adenya, a PLEC demonstration site in the forest-savanna zone of southern Ghana (Maps B and C)
- investigate the scientific basis of smallholder farmers' traditional agroforestry practices involving tree/crop combinations and the *proka* system
- carry out relevant soil nutrient analyses.

This chapter discusses the findings of that study.

Study site

The quadrangular area delimited by Obom, Asasekokoo, Kwamoso, and Mampong Nkwanta in the Gyamfiase-Adenya PLEC demonstration site (Map C) was selected for the study. It lies within the southern sector of the forest-savanna transition zone of Ghana (Hall and Swaine, 1976).

Methodology

Preliminary survey

An initial reconnaissance survey served as a basis for dividing the study area into four cells. Tree species left on farmlands in each of these cells were identified and inventoried.

Materials and method

Towards a better understanding of the efficiency of tree/crop combinations in traditional agroforestry, growth and yield studies were conducted into the following:

- maize under *Cordia millenii*
- cassava under *Ceiba pentandra*
- cassava under *Cola millenii*
- cocoyam under *Cola millenii*.

In each case, concentric rings were drawn at two-metre intervals from the bole of the selected tree. Five crops in each concentric ring were tagged and measured. The parameters considered for measurement were leaf length (l), leaf width (at $0.5l$) and numbers, plant height (h), diameter of crop (at $0.5h$), and crop yield.

For the *proka* study, one half of the field was slashed and burnt, while the other half, after slashing, was left unburnt with the slashed materials left in place as mulch. Maize was planted on both the slashed-burnt and slashed but unburnt fields. Weeding in the two fields was done after three weeks.

Soil sampling

Five soil samples were collected in each field from the four different corners and centre of the field. At each collection point, soil samples were taken at five different depths: 0–5 cm; 5–10 cm; 10–15 cm; 15–20 cm; 20–30 cm. Soils of corresponding depths were bulked.

Soil nutrient analysis

Analyses of the soil nutrient content of both slashed and burnt and slashed unburnt fields were carried out at intervals of four weeks for pH, available nitrogen, potassium, calcium, phosphorus, and organic matter content.

Soil pH was measured potentiometrically and organic matter was determined by the wet oxidation method (Wallkley and Black, 1934). Total nitrogen was determined by the Kjeldahl method, available phosphorus by the Bray and Kurtz No. 1 method, and potassium levels by flame photometry.

An analysis of variance using stratigraphics was subsequently carried out to determine the effects of:

- different tree/crop combinations and varying distance from tree trunk
- slash-and-burn and slash-without-burning practices on the growth and yield of experimental crops.

Results and discussion

Preliminary survey

Tables 14.1 and 14.2 show lists of tree and non-tree (i.e. forbes, grasses, herbs, and shrubs) species inventoried on farmer' fields in Gyamfiase-Adenya. Twenty-seven plant species belonging to 13 different plant families were encountered, with the *Sapindaceae* recording the highest number of plant species. The *Moraceae*, *Papilonaceae*, and *Apocynanceae* each had three plant species. The families *Mimosaceae*, *Asteraceae*, *Rubiaceae*, and *Boraginaceae* were each represented by only one species.

Table 14.1 Trees left *in situ* or regenerating on farms in the study area

Species	Local name	Family	Status: Left *in situ* (Is) or regeneration (R)
Albizia zygia	*Okuro*	*Mimosaceae*	Is
Alchornia cordifolia	*Agyama*	*Euphorbiaceae*	Is
Anacardium occidentale	*Atea*	*Anacardiaceae*	R
Antiaris toxicaria	*Ofun*	*Moraceae*	Is
Baphia nitida	*Odwen (barima)*	*Papilionaceae*	R
Baphia pubescence	*Odwen (obaa)*	*Papilionaceae*	R
Blighia sapida	*Ankye*	*Sapindaceae*	R
Blighia welwitschii	*Akyekobiri*	*Sapindaceae*	R
Bombax buonopozense	*Okudondo*	*Bombacaceae*	Is
Ceiba pentandra	*Onyaa*	*Bombacaceae*	Is
Cola millenii	*Wataku*	*Sterculiaceae*	Is
Cordia millenii	*Chinidro*	*Boraginaceae*	Is
Deinbolia pinnata	*Mmata*	*Sapindaceae*	R
Ficus sp.	*Nyankyeren*	*Moraceae*	Is
Funtumia elastica	*Ofuntum*	*Apocynaceae*	R
Holarrhaena floribunda	*Osese*	*Apocynaceae*	R
Lecaniodiscus cupanoides	*Odwinyinaa*	*Sapindaceae*	R
Mallotus oppositifolus	*Satadua*	*Euphorbiaceae*	R
Mangifera indica	*Mango*	*Anacardiaceae*	R
Milicia exelsa	*Odum*	*Moraceae*	Is
Milletia thoningii	*Santewa*	*Papilionaceae*	Is
Morinda lucida	*Nkyininingo*	*Rubiaceae*	R
Newbouldia laevis	*Sesrema*	*Bignoniaceae*	R
Rauvolfia vormitoria	*Akakapenpen*	*Apocynaceae*	R
Spathodea campanulata	*Bronyadua*	*Bignoniaceae*	R
Sterculia trigacantha	*Ofoso*	*Sterculiaceae*	Is
Vernonia colorata	*Anwonywini*	*Asteraceae*	R

Table 14.2 Relative abundance of non-tree germinating species

Species	Family	No-shade effect (%)	Shade effect (%)
Cyperus rotundus	*Cyperaceae*	20.2	18.5
Talinum triangulare	*Portulacaceae*	12.4	12.3
Panicum maximum	*Poaceae*	11.4	10.3
Chromolaena odorata	*Asteraceae*	10.6	8.2
Fluergya aestuans	*Urticaceae*	7.0	8.0
Euphorbia heterophylla	*Euphorbiaceae*	7.0	7.8
Setaria barbata	*Poaceae*	6.2	6.2
Phyllanthus amarus	*Euphorbiaceae*	4.2	5.1
Physalis anguiculata	*Solanaceae*	4.2	3.9
Commelina africana	*Commelinaceae*	3.9	3.6
Ageratum conyzoides	*Asteraceae*	0.0	3.1
Tridax procumbens	*Asteraceae*	2.6	2.7
Sporobulus pyramidalis	*Poaceae*	2.6	2.6
Desmodium trifoliata	*Papilionaceae*	2.1	2.1
Synedrella nodiflora	*Asteraceae*	2.1	2.1
Tragia vogelia	*Euphorbiaceae*	2.1	1.8
Emilia sonchifolia	*Asteraceae*	1.4	1.5
Xanthosoma maffafa	*Araceae*	0.0	0.2

For the non-tree plants that sprouted with food crops, there were 18 different species characterized into various life-forms that belong to 10 plant families (Table 14.2)

From their experience, traditional smallholder farmers seem to select tree species to be retained on farms irrespective of plant families. However, the choice of tree species to be left on farms may be dependent on the contribution that tree can make to crop growth and yield, since no one particular family is overwhelmingly present on the farms. The criteria for selection may be the nature of the litter produced and the ease or rapidity with which it can decompose to add organic matter to the soil, although these particular parameters were not investigated in the study.

Effects of types of tree and distance from tree trunk on height/growth of crops

Forbe (non-tree) species that were found germinating in the concentric rings closer to the trunks or boles of the experimental trees (shade effects) were the same as those observed in the concentric rings farther away from the boles of the experimental trees (no-shade effects; Table 14.2). No significant difference was noted in the percentage germination of each of these species under both conditions, i.e. either close to the boles (shade effects) or further away (no-shade effects; Table 14.2).

Generally, yield and height of crops appeared to increase with increasing distance from the trees irrespective of the crop type (Table 14.3; Figures 14.1–14.3), except for cocoyam growing under *Cola millenii* (Figure 14.4) where plants growing two metres from the bole were taller than those growing 10 m from the bole or tree trunk.

For cassava and maize, yield increased as distance from tree trunk is increased, with *Ceiba pentandra* having a greater stimulating influence (on the yield) than *Cola millenii*. However, at the time of preparing this chapter, the yield of cocoyam growing under *Cola millenii* had not been determined.

From the results, it appears that shading contributes to reduced growth vigour of crops. The only exception is cocoyam, which tended to do better closest to the bole of *Cola millenii*, which, according to farmers, generally combines poorly with other crops.

Table 14.3 Crop yields (kg) in relation to distance from the trunk/bole of tree

Crop under different tree combinations	Distance from bole/trunk of tree				
	2 m	4 m	6 m	8 m	10 m
Cassava under	1.80	7.80	13.80	14.00	17.50
Ceiba pentandra	(0.30)	(1.30)	(2.30)	(2.33)	(2.92)
Cassava under	1.80	3.90	9.60	11.40	14.0
Cola millenii	(0.30)	(0.65)	(1.60)	(1.90)	(2.33)
Maize under	0.60	0.65	0.65	0.70	0.75
Cordia millenii	(0.12)	(0.13)	(0.13)	(0.14)	(0.15)

Notes:
1. Crop planted in concentric rings two metres apart from base of tree bole/trunk.
2. Upper values represent total yield per concentric ring.
3. Lower values in parenthesis represent mean weight per crop.

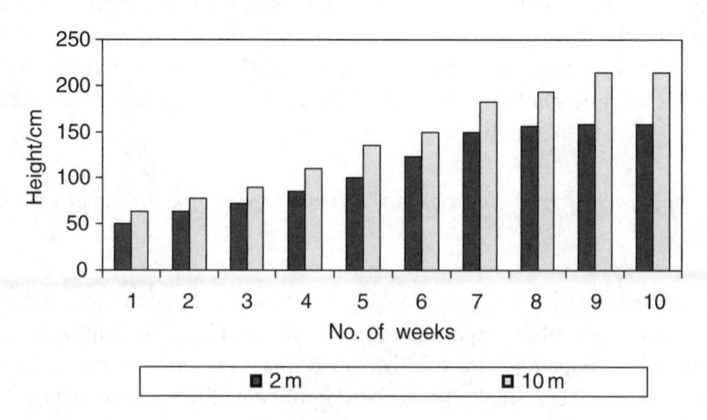

Figure 14.1 Effect of distance from tree trunk (2 m or 10 m) on height growth of maize under *Cordia millenii*

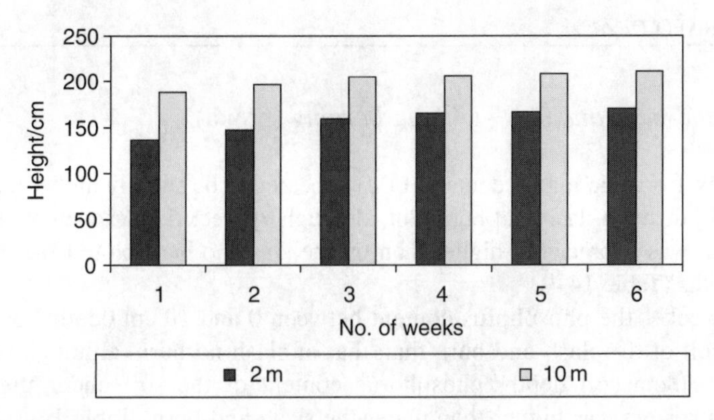

Figure 14.2 Effect of distance from tree trunk (2 or 10 m) on growth in height of cassava plants under *Ceiba pentandra*

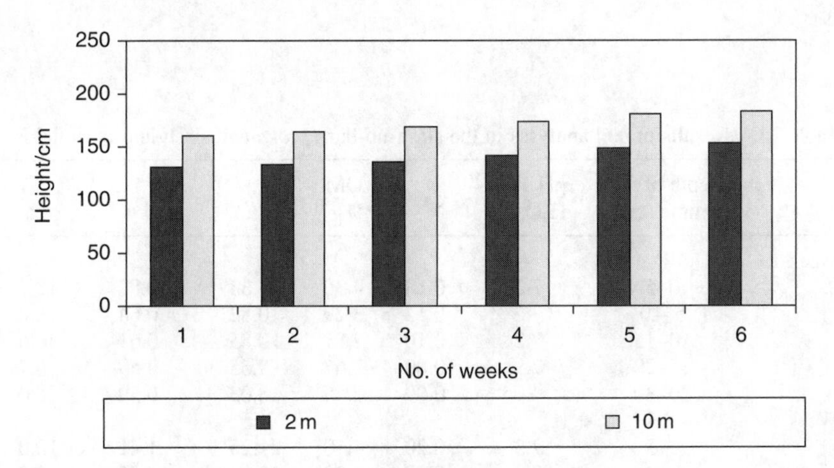

Figure 14.3 Effect of distance from tree trunk (2 m or 10 m) on growth in height of cassava plants under *Cola millenii*

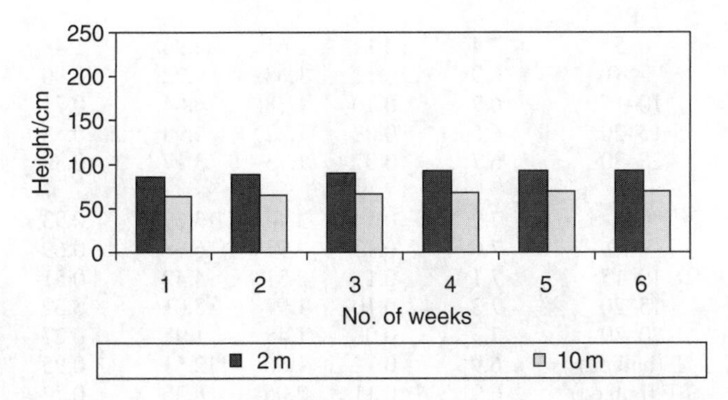

Figure 14.4 Effect of distance from tree trunk (2 m or 10 m) on height of cocoyam under *Cola millenii*

Slash-and-burn and slash without burning (proka)

For week 1 organic matter content (OM) appeared to be slightly higher under the *proka* (P) than the slash-and-burn plot, although by week 5, OM content on slash-and-burn was appreciably higher than on the slash-no-burn plot at the different soil depths (Table 14.4).

For week 1 the phosphorus content between 0 and 10 cm depth was higher in the soil of the slash-and-burn than that of slash-no-burn, although between 10 and 30 cm soil depth, phosphorus content of the soil under the slash-no-burn system was higher than under the slash-and-burn (Table 14.4).

By week 5 the phosphorus content in the soil under the slash-no-burn system was higher at all depths when compared to the soil of the slash-and-burn.

Table 14.4 Results of soil analyses in the slash-no-burn *proka* and slash-and-burn fields

	Depth of sample (cm)	pH 1:1 H_2O	N	OM %	P	K mg kg^{-1}	Ca
Week 1	P						
	0–5	7.3	0.18	4.30	29.33	0.82	13.8
	5–10	7.2	0.12	3.29	10.82	0.64	8.0
	10–15	7.2	0.10	2.81	10.85	0.61	6.6
	15–20	7.1	0.09	2.67	9.65	0.59	6.2
	20–30	6.9	0.09	2.65	8.03	0.59	6.0
Week 1	BF						
	0–5	7.3	0.20	4.10	31.25	1.41	10.0
	5–10	7.4	0.13	3.22	13.78	0.95	9.8
	10–15	7.1	0.13	2.81	8.30	0.72	8.2
	15–20	7.2	0.10	2.55	5.63	0.69	7.6
	20–30	7.4	0.08	2.10	5.18	0.54	6.4
Week 5	P						
	0–5	7.4	0.14	2.61	34.96	1.48	8.6
	5–10	6.9	0.11	1.68	9.22	0.96	5.8
	10–15	6.7	0.10	1.38	6.64	0.71	5.2
	15–20	6.6	0.08	1.22	4.98	0.58	4.0
	20–30	6.7	0.07	1.13	3.17	0.39	3.8
Week 5	BF						
	0–5	7.2	0.14	3.44	13.69	0.93	10.4
	5–10	7.0	0.12	2.97	6.98	0.69	8.6
	10–15	7.1	0.11	2.31	4.42	0.51	7.0
	15–20	7.3	0.10	1.97	3.01	0.32	7.2
	20–30	7.3	0.08	1.48	1.95	0.27	6.6
Week 5	P (bulked)	6.9	0.12	1.66	12.61	0.95	5.8
	BF (bulked)	6.9	0.11	2.06	6.36	0.72	7.0

OM – Organic matter; P – *Proka* field; BF – Burnt field.

Generally the potassium at virtually all depths was higher in the soil under the burnt conditions than under the unburnt conditions at week 1. By week 5, however, there was a higher potassium content at all the different depths of the soil under the slash-no-burn conditions.

Except for soils from 0 to 5 cm during week 1, the calcium content at all depths was relatively higher under the burnt conditions when compared to the slash-no-burn conditions for both weeks 1 and 5 (Table 14.4).

In general, there was no significant difference between all the vegetative growth parameters of the maize crop that were studied under both the slash-and-burn and slash-without-burning (*proka*) conditions (Figures 14.5–14.9). This is evident in the height growth (Fig. 14.5). Similar observations were made for the leaf length (Figure 14.6) number of leaves (Figure 14.7), leaf width (Figure 14.8) and mean stem diameter (Figure 14.9).

In terms of yield, each crop produced a maize cob irrespective of whether it was from a slash-and-burn or slash-no-burn *proka* plot. The weights or biomass of the entire crop (not dehusked) were 2.10 kg and 2.40 kg respectively for the plants under both *proka* and slash-and-burn experimental conditions. Crop yield was slightly higher on the slash-and-burn field than in the slash-no-burn or *proka* field. This small difference may be reflective of the fertilizer effect of the ash when compared to the situation under the *proka* farm, which has no ash associated with it because of the avoidance of burning. Total yield of maize produced under the slash-and-burn condition was slightly higher (2.40 kg) than that under the slash-but-unburnt (*proka*) condition (2.10 kg). However, statistically, these values were not found to be significant (Table 14.5; Figure 14.5).

The results obtained seem to indicate that the *proka* method may not confer any advantage to the growth and development of crops in a slash-no-burn

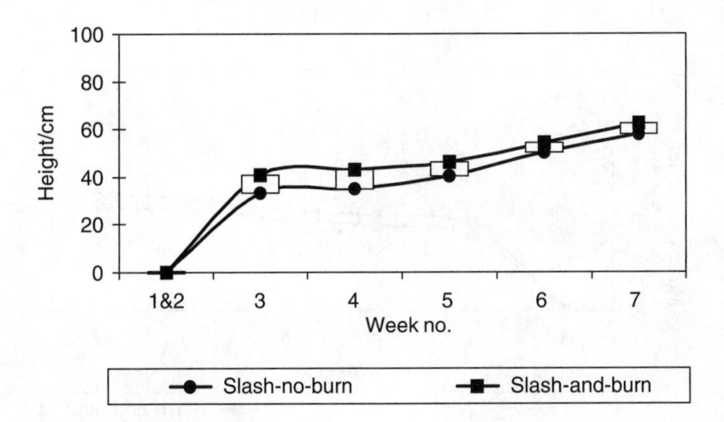

Figure 14.5 Effect of slash-and-burn and slash without burning on vegetative growth of maize

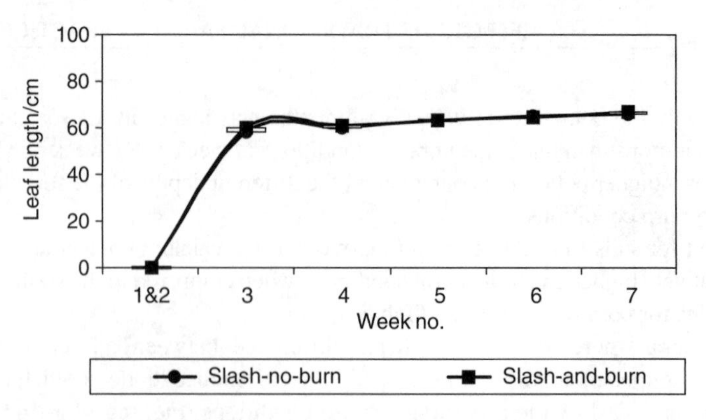

Figure 14.6 Effects of slash-and-burn and slash without burning on leaf length on maize

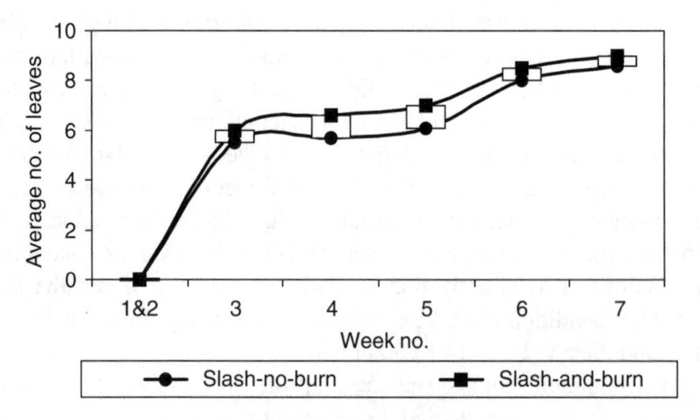

Figure 14.7 Effects of slash-and-burn and slash without burning on number of leaves of maize

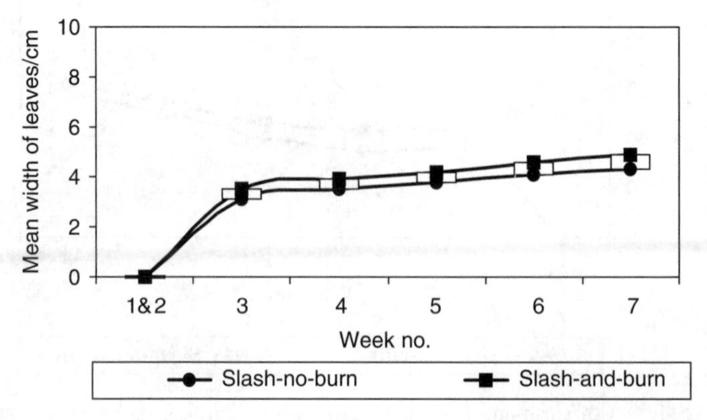

Figure 14.8 Effects of slash-and-burn and slash without burning on leaf width of maize

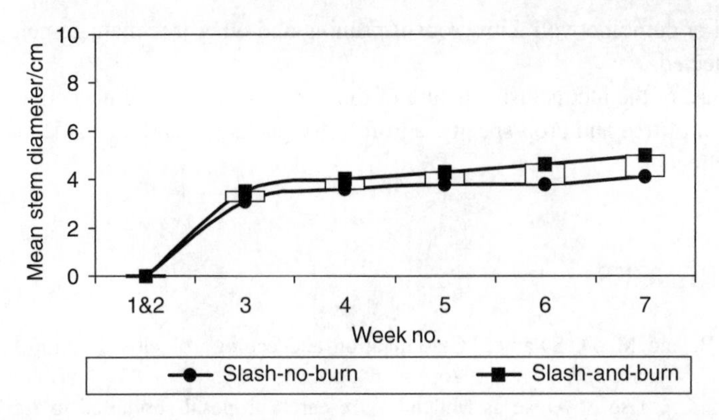

Figure 14.9 Effects of slash-and-burn and slash without burning on mean stem diameter of maize

Table 14.5 Total yield (kg) of maize obtained from *proka* and slash-and-burn fields

Farm practice	Total yield/kg	Mean weight per cob
Slash-no-burn (*proka*)	2.10	0.13
Slash-and-burn	2.40	0.15

Total represents weight of entire maize harvest.

plot when compared to the slash-and-burn plot where crops seem to have performed more robustly in terms of vegetative vigour and yield (Table 14.5). This finding is in sharp contrast to the finding reported for a PLEC demonstration site in the humid forest zone of central Ghana by Quansah and Oduro in Chapter 11.

However, it is likely that, eventually, the *proka* system would contribute positively to crop growth and yield, since the release of mineral nutrients from decomposed litter is a slow process (Markwei, 2002).

Conclusion

The preceding discussion suggests that the agronomic requirements of individual crop species may be important when considering the combination of trees and crops on farms, and that management of such trees in terms of pruning to eliminate shade may be very important in designing agroforestry packages.

It is therefore likely that in general, field crops (e.g. corn and cassava) that require higher light intensity may do relatively poorly even under trees that are known to be good or combine well with crops if pruning and other tree management aspects are neglected.

Because of the inconclusive nature of the findings, studies are needed, especially on design of tree and crop-specific agroforestry packages, and on shade and allelopathic effects.

REFERENCES

Hall, J. B. and M. D. Swaine, "Classification and ecology of closed canopy forest in Ghana", *Journal of Ecology*, 1976, Vol. 64, pp. 913–951.

Markwei, C., "Use of Algae as Mulch", A research proposal submitted to Agricultural Services Sub-Sector Investment Programme (AgSSIP), January (2002), unpublished.

Wallkley, A. and I. A. Black, "An examination of the Degtareff method for determining soil organic matter and a proposed modification of the chromic acid titration method", *Soil Science*, Vol. 37, 1934, pp. 29–38.

15

Effects of four indigenous trees canopy covers on soil fertility in a Ghanaian savanna

Charles Anane-Sakyi, Abubakari Sadik Abdulai, and Saa Dittoh

Introduction

The northern savanna zone covers approximately two-thirds of the total area of Ghana (Map B). It is mostly affected by different seasons, especially the *harmattan* dry season. Its natural climax vegetation is characterized by a ground flora of perennial grasses like *Panicum maximum*, *Pennisetum purpureum*, *Andropogon gyanus*, and *P. pedicelatum*, with scattered, more or less fire-resistant deciduous broad-leafed trees and shrubs of various sizes and densities. In the extreme north-eastern area of the Sudan savanna, the common tree species are *Vitellaria paradoxa* (sheanut tree), *Parkia biglobosa* (*dawadawa* tree), *Faidhebia albida* (*Acacia albida*), *Adansonia digitata* (baobab), and *Diospyrum mespiliformis* (ebony tree). They are grown in an orchard pattern with food crops underneath even in densely cropped areas.

Owing to pressure on the land caused by high human population density, many farmers remove the indigenous trees to increase cropping space without considering the effect of such a practice on the soil properties, especially soil fertility. Moreover, various agroforestry systems are being promoted in the area with exotic woody species without considering the effects on the indigenous trees. There is abundant information on the influence of trees and shrubs on soil fertility in a wide variety of ecosystems (Radwanski and Wicken, 1967; Gerakis and Tsangarakis, 1970; Kellman, 1979; Hattan and Smart, 1984; Belsky et al., 1989; Isichei and Muoihalu, 1992) which indicates that woody species increase the fertility of the soil under their canopies.

In northern Ghana, especially in the north-eastern zone, there is a lack of information on the relationship between the indigenous tree canopies and soil fertility. This prompted a study in 1995 and 1996 to evaluate the effects of the canopies of four indigenous woody species on savanna soil properties. This chapter reports the findings of that study, which was carried out under the United Nations University Project on People, Land Management, and Environmental Change (UNU/PLEC).

Study area

The site lies in the natural Sudan savanna in the Upper East region (UER; Map B). It is bounded by Bawku, Binduri, and Bugri in the Bawku-East district, and drained by the River Tamne, a tributary of the White Volta River that has many small dams and river valleys. The upland soils are mainly luvisol, lixisol, plinthosol, and fluvisol derived from biotite granite. These have sandy topsoils, sandy clay subsoils, and low organic matter. The valley bottom soils are gleysol derived from basic metamorphites (mainly greenstone). They are clayey and rich in organic matter with high stone content, which increases down to the subsoil (leptosol and cambisol; Hauffe and Jaensch, 1992). The soil/moisture relationship is extreme. Annual mean rainfall varies from 645 mm to 1,250 mm with a long mean of 1,044 mm. Rainfall is limited to five months of the year. The remaining months are mostly dry. Temperatures range from an annual mean minimum of 22.5°C to a mean maximum of 35.6°C. Phosphorus and nitrogen are scarce (Adu, 1969; Boateng and Ayamga, 1992).

Materials and methods

Four naturally occurring indigenous trees, *F. albida* (*A. albida*), *D. mespiliformis* (ebony tree), *V. paradoxa* (sheanut tree), and *P. biglobosa* (*dawadawa* tree), on two different soil types (upland soil, luvisol; valley soil, gleysol) were studied. These trees were chosen because of their occurrence on farmlands and their importance to the communities, as shown in Table 15.1.

The data in Table 15.1 were obtained in discussions with the farmers during transect walks across the various topographic lines in the study area. The soils were sampled from under the tree canopies and from open grassland. At each site under the tree canopies, samples were taken from four different spots and hand mixed together to give a bulk sample (about 1.5 m from the tree trunk). Soil samples were also taken from the open grassland. A total of 48 samples each were collected on luvisol and gleysol. All samples were taken at a depth of 15 cm from cultivated soils. Leaf samples from the trees were

Table 15.1 Four indigenous tree species and their socio-economic importance in the north-eastern savanna zone of Ghana

	Food				Fodder	Wood				
Species	Fruits/nuts	Oils and fats	Spices or flavours	Beverages		Firewood	Charcoal	Building material	Utensils and tools	Timber
Faidhebia albida	X				X	X	X	X	X	X
Vitellaria paradoxa	X	X			X	X	X	X	X	
Parkia biglobosa	X		X	X	X	X				X
Diospyrum mespiliformis	X				X	X	X	X	X	X
	Services					Miscellaneous				
	Shade	Soil conservation	N-fixing	Tannins		Dyes	Fibres	Medicine		
Faidhebia albida	X	X	X	X				X		
Vitellaria paradoxa	X							X		
Parkia biglobosa	X	X		X		X	X	X		
Diospyrum mespiliformis	X							X		

also taken. Soil samples taken were lightly crushed and passed through a 2 mm sieve. The leaf samples were oven-dried at 70–80°C and finely crushed. Soil pH, organic carbon, total nitrogen, available phosphorus, exchangeable cations, and cation exchange capacity were determined. Soil pH was determined in INKCI at a soil:solution ratio of 1:2. Organic matter was determined by the Walkey-Black method (Allison, 1965). Total nitrogen was determined by the Kjeldahl digestion method (Bremmer, 1965). Exchangeable cations (Ca, Mg, K, Na) were extracted by neutral 1 M ammonium acetate solution. Potassium and sodium in solution were estimated by flame emission photometry, and calcium and magnesium by atomic absorption spectrophotometry. Available phosphorus was determined by the Bray 1 method (Bray and Kurtz, 1945) and cation exchange capacity (CEC) by ammonium acetate extraction at pH 7.0 (Pleysier, 1982). Base saturation was calculated as the sum of total exchangeable bases as a percentage of CEC. The values of the chemical properties of the two soil types under the tree canopies and in open grassland were pooled to assess the effect of the indigenous trees on the soil chemical properties.

Results

Socio-economic importance of indigenous trees

From Table 15.1, one realizes that all the four indigenous trees have complementary uses. With respect to soil conservation, however, it is *F. albida* and *P. biglobosa* that are important. All are important as fodder for animals and as fuelwood. Even though farmers recognize the importance of all these trees, they still regard them as hindrances and, therefore, cut them out of their farms to increase the area for arable crops.

Analysis of leaf samples

Of all the trees, *F. albida* leaves had the highest total nitrogen, phosphorus, and potassium contents, as shown in Table 15.2. There is an indication of the usefulness of *F. albida* leaves in improving soil fertility and as fodder for animals.

Among the four indigenous trees, the chemical properties of both soil types under their canopies were highest in *F. albida*, followed by *D. mespiliformis*, *P. biglobosa*, and *V. paradoxa* in that order. Significant difference in soil

Table 15.2 Nitrogen, phosphorus, and potassium contents of the leaves of the four indigenous trees

Nutrient content	*Faidhebia albida*	*Diospyrum mespiliformis*	*Vitellaria paradoxa*	*Parkia biglobosa*
Total nitrogen %	2.85	1.45	1.75	1.70
Total phosphorus %	0.60	0.45	0.55	0.44
Potassium (mg/kg)	1,152.0	1,132.0	688.0	635.0
N/P ratio	4.75	3.22	3.20	3.95

properties in between tree canopies and the open grassland for both soil types were largest in *D. mespiliformis*, followed by *P. biglobosa*, *F. albida*, and *V. paradoxa* in that order.

Soil chemical properties under tree canopies and in open grassland

Comparison of chemical properties of soils under tree canopies and under open grasslands was initially done for each indigenous tree under luvisol and gleysol separately. In all cases, all chemical properties except C:N ratio were higher under tree canopies than in open grassland. When the data were put together to provide an overall comparison under luvisol (Table 15.3) and gleysol (Table 15.4), soil pH values under all trees for both soil types showed that the soils were weakly acidic (pH 5–6). Soil pH values under tree canopies were significantly higher than in the open grassland in luvisol (Table 15.3), but not significant in the gleysol (Table 15.4). Soil organic matter and total nitrogen content under tree canopies were significantly higher than in the open grassland in both soil types. There was no significant difference in the carbon:nitrogen (C:N) ratio between the two soil types under tree canopies or in the open grassland. There was no significant difference in available soil phosphorus in the gleysol between the tree canopies and the open grassland. However, the differences were significant in the luvisol. The content of soil exchangeable cations, calcium, magnesium, potassium, and sodium in both soil types under the tree canopies was higher than in the open grassland (Tables 15.3 and 15.4). It was significant for all elements on both soil types except potassium in gleysol (Table 15.4). The total exchangeable bases content followed the same trend as the individual cations. It was higher under tree canopies than in the open grassland, and the difference was significant for all trees in both soil types. Cation exchange capacities under tree canopies were higher than in the open grassland for both soil types.

Table 15.3 Mean values ± 90 per cent confidence interval of soil chemical properties under four indigenous trees canopies and in the open grassland on luvisol in the north-east savanna zone of Ghana

	Faidhebia albida		*Diospyrum mespiliformis*		*Vitellaria paradoxa*		*Parkia biglobosa*	
Soil property	Open	Tree canopy	Open	Tree canopy	Open	Tree canopy	Open	Tree canopy
PH (KCI)	6.00	6.00	5.00	6.00	5.00	5.00	5.00	6.00*
% organic matter	0.79	1.88*	0.80	1.21*	0.66	1.20*	0.52	1.12*
% total nitrogen	0.04	0.09*	0.03	0.06*	0.03	0.05*	0.02	0.05*
C:N ratio	14.00	12.00	13.00	12.00	17.00	14.00	15.00	12.00
Available P (mg kg^{-1})	0.81	2.61*	1.00	2.30*	1.34	1.72*	0.48	1.68*
Exchange cations (cmol kg^{-1})								
Calcium	2.27	3.31*	3.02	4.50*	1.84	2.73	2.80	4.23*
Magnesium	1.37	1.78	1.75	2.72*	1.40	1.77	1.53	2.75*
Potassium	0.12	0.19*	0.12	0.15	0.08	0.17*	0.11	0.23*
Sodium	0.27	0.49*	0.22	0.67*	0.22	0.39*	0.24	0.48*
Total exchangeable bases (cmol kg^{-1})	4.03	5.59	5.11	8.04*	3.56	5.06	4.35	7.52*
Cation exchange Capacity (cmol kg^{-1})	5.12	6.50	6.26	8.49*	4.64	5.86	5.65	9.11*
% base saturation	78.98	84.52	81.45	94.95*	76.17	87.09*	77.43	81.95

* Differences significant at $P < 0.05$ between values for canopy and open lands.

Table 15.4 Mean values ± 90 per cent confidence interval of soil chemical properties under four indigenous trees canopies and in the open grassland on gleysol in the north-east savanna zone of Ghana

	Faidhebia albida		*Diospyrum mespiliformis*		*Vitellaria paradoxa*		*Parkia biglobosa*	
Soil property	Open	Canopy	Open	Canopy	Open	Canopy	Open	Canopy
pH (KCI)	6.00	6.00	6.00	6.00	5.45	5.67	5.00	5.00
% organic matter	1.33	2.42*	0.50	1.20*	0.87	1.31*	0.99	1.47*
% total nitrogen	0.06	0.11*	0.03	0.07*	0.03	0.05*	0.04	0.07*
C:N ratio	13.00	12.00	15.00	12.00	17.00	15.00	14.00	12.00
Available P (mg kg^{-1})	0.54	2.10	7.30	8.54	1.25	1.59	0.45	1.15
Exchange cations (cmol kg^{-1})								
Calcium	2.20	4.20*	2.37	4.30*	2.53	3.37	2.07	4.40*
Magnesium	1.40	2.80	1.67	2.47	1.78	2.77	1.43	2.38
Potassium	1.45	0.23	0.12	0.21	0.09	0.23	0.10	0.17
Sodium	0.45	0.62*	0.28	0.61*	0.18	0.33*	0.23	0.42*
Total exchangeable bases (cmol kg^{-1})	4.17	7.71*	4.55	7.59*	4.84	5.84	3.84	7.37*
Cation exchange Capacity (cmol kg^{-1})	6.11	9.02*	6.55	8.58*	6.12	6.82	5.12	8.40*
% base saturation	74.00	83.31	65.73	86.47*	76.67	85.80	73.81	88.11*

* Differences significant at $P < 0.05$ between values for canopy and open lands.

Discussion

A number of studies have reported higher values for soil chemical properties under tropical tree canopies (Kellman, 1979; Hattan and Smart, 1984; Belsky et al., 1989; Isichei and Muoihalu, 1992) than in the open grassland, as found in this study. Because of the accumulation of leaf fall, the higher organic matter content beneath tree canopies reduces leaching, slows the rate of mineralization as a result of reduction in temperature, and induces a more favourable microclimate under the trees. Other advantages include deposition of manure and urine there by livestock seeking shade or food, nitrogen fixation by the trees (*F. albida*), and recycling by their deep roots of nutrients from lower soil levels. The high total nitrogen content under tree canopies is also likely due to the high nitrogen content of their leaves as compared to the poor nitrogen content of the natural grass vegetation in grass-dominated savanna soils (Jones, 1973).

As there is a direct correlation between soil organic matter level and cation exchange capacity, it was not surprising that the cation exchange capacity of the soils under the tree canopies was higher than in the open grassland. Among the tree species the high values of chemical properties of soil under *F. albida* are due to the relatively high nutrient content of its leaves. The low values of N:P ratio could not have been accounted for by the immobilization of the N and P under tree canopies. There was no significant difference in available soil phosphorus for the gleysol between tree canopies and open grassland. This is because of the higher clay content of gleysol, which prevents leaching in both circumstances, and also the low crop removals of phosphorus from the soil.

Among the two soil types, the soil chemical properties in gleysol were higher than in luvisol. This might be due to the difference in texture. It has been reported that in West African savanna (Jones, 1973) and in East Africa, soil texture is the most important factor controlling organic matter and, consequently, other soil chemical properties. Differences may also be due to the parent rock material and topography since luvisol are mostly upland and gleysol are in valleys. Moisture and other relations could also be the result of their stated differences as well.

Policy implication findings

The results of the study show that indigenous savanna trees have a beneficial effect on soil nutrient status and, therefore, on soil productivity in Ghanaian savanna soils. Accordingly farmers need to be encouraged to cultivate and maintain these trees on their farms. Also there is need to plant indigenous trees in agroforestry systems. It is surprising that indigenous trees have been largely ignored in agroforestry systems in the savanna areas of Ghana.

Small-scale livestock farmers have known that a number of indigenous trees are good sources of fodder, but animal scientists have given little attention to these trees as sources of fodder. There is need to correct this situation as more and more attention is being given to zero-grazing and nutrient-recycling production systems. In a free-range livestock system great heed should be given to these indigenous trees as they are not destroyed by the livestock and are moreover fire resistant.

The four indigenous trees studied must be considered for use in improving and maintaining the fertility status and hence the productivity of Ghanaian savanna soils.

Acknowledgements

The authors are grateful to the UNU/PLEC project for funding the research from which this chapter is derived. They also wish to express appreciation to Geoffrey Affi-Punguh and Abdulai Abo, both technical staff of Manga Research Station, for their assistance during the fieldwork, and to Professor A. B. Ankomah of the Department of Crop Science, Kwame Nkrumah University of Science and Technology, for his useful suggestions and comments.

REFERENCES

Adu, S. V., *Soils of the Navrongo Bawku Area Upper Region Ghana*, Soil Research Institute Memoir, No. 5, Kumasi, 1969.

Allison, L. E., "Organic matter PP 1367–1368, in Black C. A method of soil analysis par 2. Chemical and microbiological properties", in *Agronomy 9*, Madison, WI: American Society of Agronomy 1965.

Belsky, A. J., R. G. Amundson, J. M. Duxbury, S. J. Riha, A. R. Ail, and S. M. Mwonga, "The effects of trees on their physical, chemical and biological environment in a semi-arid savanna in Kenya", *Journal of Applied Ecology*, Vol. 26, 1989, pp. 156–167.

Boateng, E. and T. Ayamga, "Soil and Land Evaluation Studies at Tono-Navrongo, Wiaga, Zuanrungu and Manga Agricultural Station, Upper East Region, Ghana", IFAD/MOALACOSREP (UER), 1992.

Bray, R. H. and L. T. Kurtz, "Determination of total organic and available forms of phosphorus in soil", *Soil Science*, Vol. 59, 1945, pp. 39–45.

Bremmer, J. M., "Determination of nitrogen in soil by the Kjedhal method", *Journal of Agricultural Science*, Vol. 55, 1965, pp. 1–23.

Gerakis, P. A. and C. Z. Tsangarakis, "The influence of *Acacia Senegal* on the fertility of a sand sheet (Goz) soil in the central Sudan", *Plant and Soil*, Vol. 33, 1970, pp. 81–86.

Hattan, J. C. and N. D. E. Smart, "The effect of status in Murchson Falls, National Fall, Uganda", *African Journal of Ecology*, Vol. 22, 1984, pp. 23–30.

Hauffe, H. K. and S. Jaensch, "Development and properties of some soils derived from granite and basic metamorphites in the Upper East region of Ghana", in *Proceedings of the 12th and 13th Annual General Meetings of the Soil Science Society of Ghana*, 1992, pp. 103–108.

Isichei, A. O. and J. I. Muoihalu, "The effects of tree canopy on soil fertility in a Nigerian savanna", *Journal of Tropical Ecology*, Vol. 8, 1992, pp. 329–338.

Jones, M., "The organic matter contents of the savannah soils in West Africa", *Journal of Soil Science*, Vol. 24, 1973, pp. 42–53.

Kellman, M., "Soil enrichment by neotropical savanna trees", *Journal of Ecology*, Vol. 67, 1979, pp. 567–577.

Pleysier, J. L., *Course in Soil and Plant Analysis Lecture Notes*, Ibadan: IITA, 1982.

Radwanski, S. A. and G. E. Wicken, "The ecology of *Acacia albida* on mantle soils in Zalinger, Jebel Marra, Sudan", *Journal of Applied Ecology*, Vol. 4, 1967, pp. 567–576.

16

Comparative management of the savanna woodland in Ghana and Guinea: A preliminary analysis

Lewis Enu-Kwesi, Vincent V. Vordzogbe, Diallo Amirou, and Diallo Daouda

Introduction

The Republics of Ghana and Guinea are both located in West Africa. They have similar vegetation zones that have been modified through anthropogenic factors such as agriculture, logging, and the quest for biomass either directly as fuelwood or for conversion into charcoal.

Recently Thies (1995) classified the forest-savanna transition zone of West Africa to cover virtually the entire area between 5°N and 10°N. This broad area includes the forest-savanna transition zone as well as the entire Guinea savanna zone of Ghana and the area around Kouroussa-Moussaya in north-eastern Guinea (Figure 16.1; see also Map A; Swaine and Hall, 1986).

Preliminary observations had indicated close similarity as well as differences in the composition of species in this broadly classified transition zone. There also appears to be very close cultural similarity among people of Ghana and Guinea in terms of the use of flora to enhance their livelihoods.

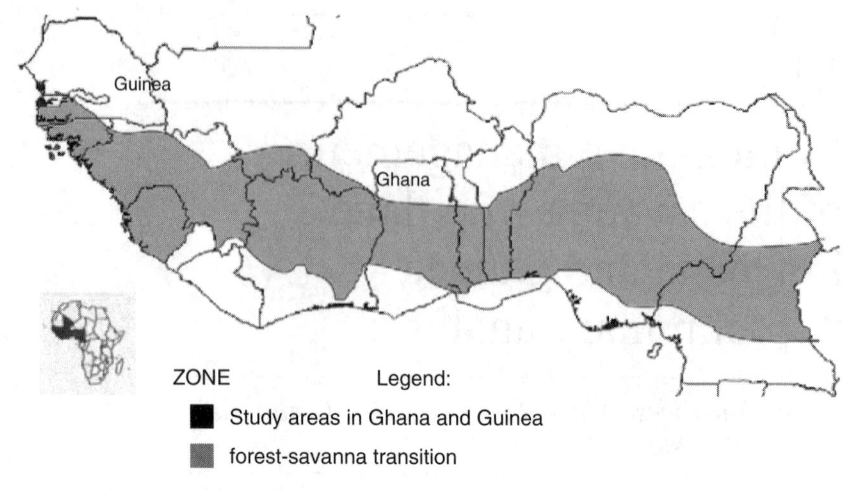

Figure 16.1 Map of West Africa showing location of the study sites in relation to the forest-savanna transition zone
Source: Thies (1995)

A comparative study was designed to investigate the impact of human use and other factors on the vegetation of these two countries.

Methodology

Study area

The chosen study sites are Kouroussa-Moussaya in Guinea, and the Asantekwa area (including Tahiru-Akuraa, Bawa-Akura, and Dawadawa) and Tamale area (including Binguri, Dugu-Song, Bognayili, and Jaagbo) in Ghana. A protected forest in the Sudan savanna of Ghana at Binguri was studied as the reference for comparison to the protected forest at Kouroussa-Moussaya.

Procedure

The PLEC-BAG and PLEC-STAT methodologies recommended by Zarin, Guo, and Enu-Kwesi (1999) and Coffey (2000) were followed to determine:

- species composition, similarity, and richness
- frequency distribution of common species in the plots of the two countries
- general rank order of species
- abundance diversity and Shannon-Weiner diversity
- girth- and height-class distribution of woody species
- utility.

Results

Species composition

In both countries, sampling proved effective and efficient as is evident in the species/area curves (Figure 16.2). Species richness in the plots studied was higher in Ghana (103) than in Guinea (82).

Altogether, 38 similar species observed in the fields of the two countries and their frequency distribution were estimated (Tables 16.1 and 16.2) and graphically presented (Figures 16.3 and 16.4). The following common woody trees occur more frequently in Guinea than in Ghana: *Afzelia africana*, *Annona senegalensis*, and *Terminalia albida*. By contrast, species such as *Vitex doniana*, *Lannea nigricans*, and *Combretum ghasalens* are higher in relative frequency of occurrence in Ghana. However, there appeared to be very little difference in the relative frequencies of the woody species *Vitellaria paradoxa*, *T. albida*, *Strychnos spinosa*, *Ficus sur*, *Cussonia djalonensis*, and *Cassia siberiana* in the plots that were inventoried in the two countries.

The cumulative frequencies of both woody and non-woody species in the plots studied were nearly similar and normally distributed (Table 16.2). However, the spread in occurrence of comparable plant species in the sample plots in the two countries show differences in shape and height models. Their histogram plots indicate quantitatively variable species occurrence. The valid frequencies fall within commonly distributed intervals with similar peak centres (Figs 16.3 and 16.4). The mean species similarity in sample plots of the two countries is estimated to be 50 ± 2 per cent (Tables 16.3 and 16.4).

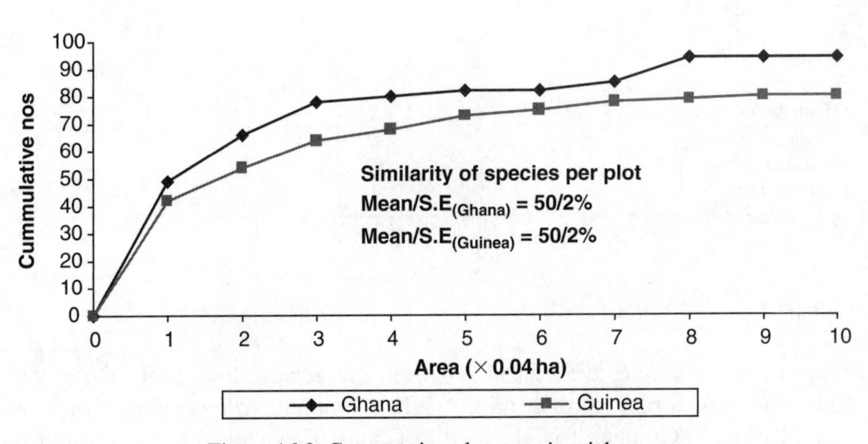

Figure 16.2 Comparative plant species richness.

Table 16.1 Frequency of comparable species prevalent in sample plots in Ghana and Guinea

Species	Ghana	Guinea
Afzelia africana	1	6
Amorphophalus sp.	1	1
Annona senegalense	2	4
Bridelia ferruginea	1	4
Cassia siberiana	1	1
Cissus sp.	3	–
Cissus aralioides	–	1
Cissus doeringii	–	1
Cochlospermum planchonnii	–	2
Cochlospermum thincktorium	1	–
Combretum ghasalens	3	1
Combretum nigricans	1	2
Commelina sp.	1	4
Cussonia djalonensis	2	1
Daniella oliveri	3	6
Desmodium giganticum	2	1
Desmodium adscendens	1	1
Ficus sur	2	1
Gardenia sp.	3	5
Khaya senegalensis	–	1
Lannea acida	–	3
Lannea macrocarpa	1	–
Lannea nigricans	3	–
Lophira lanceolata	2	5
Parkia clappertoniana	3	5
Piliostigma sp.	2	4
Strychnos densifolia	–	1
Strychnos spinosa	3	2
Tephrosia elegans	2	–
Tephrosia linearis	2	–
Tephrosia sp.	1	1
Terminalia albida	4	5
Terminalia glauscens	1	–
Trichilia emertica	3	1
Vernonia colorata	2	1
Vitellaria paradoxa	6	5
Vitex doniana	4	1

General rank order of species

When selected species are considered from the general list (Table 16.5), the following rank order is discernible. While *Daniella oliveri* ranks first in sample plots in Guinea, it is eleventh in Ghana. *Chromolaena odorata* ranks

Table 16.2 Cumulative frequency of comparable species in sample plots in Ghana and Guinea

Valid frequency	Ghana			Guinea		
	Counts	% valid	% cumulative	Counts	% valid	% cumulative
0	5	13.5	13.9	7	18.9	18.9
1	11	29.7	44.4	15	40.5	59.5
2	9	24.3	69.4	3	8.1	67.6
3	8	21.6	91.7	1	2.7	70.3
4	2	5.4	97.2	4	10.8	81.1
5	–	2.7	97.2	5	13.5	94.6
6	1	2.7	100.0	2	5.4	100.0
Total	36	100.0	–	37	100.0	–

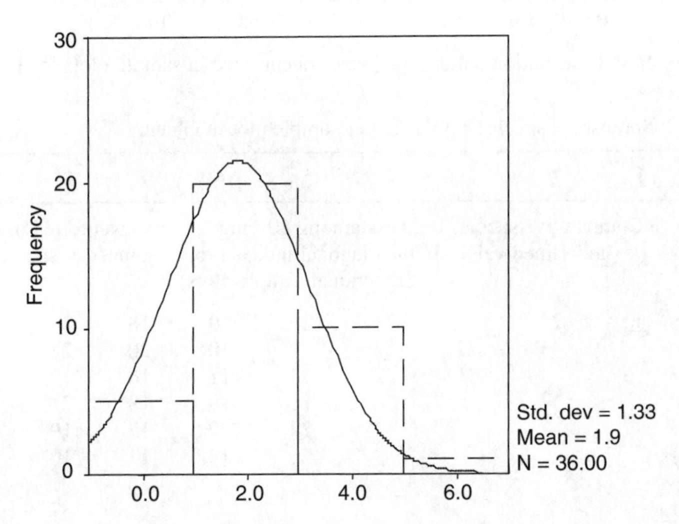

Figure 16.3 Distribution pattern of species occurrence in sample plots in Ghana

first in the plots studied in Ghana. *Lophira lanceolata* ranks thirty-third in Ghana, but fourteenth in Guinea. *Vitelleria paradoxa* ranks fourth in Ghana whereas it is twenty-fifth in Guinea. *Piliostigma* sp. is twenty-fourth in Guinea, but thirty-fourth in Ghana. However, *Parkia clappertoniana*, *Gardenia tenuifolia*, and *A. senegalensis* rank similarly in both countries.

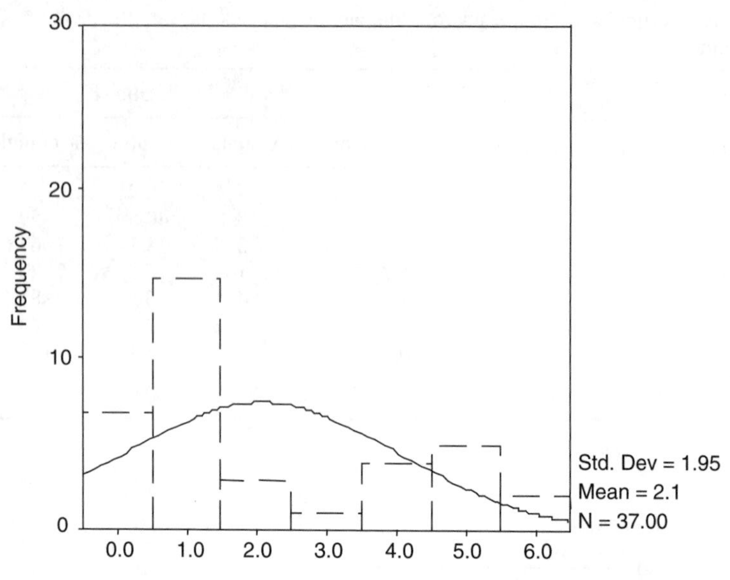

Figure 16.4 Distribution pattern of species occurrence in sample plots in Guinea

Table 16.3 Sorenson's species similarity per sample plot in Ghana

Plot	1	2	3	4	5	6	7	8	9
	Comparative species richness among 10 sample plots inventoried in Ghana (underlined values in the diagonal indicate total richness of species in individual sample plots)								
1	50	25	18	21	21	9	18	28	25
2		43	22	23	18	10	20	22	16
3			41	22	15	11	16	19	18
4				36	18	10	18	21	21
5					28	9	17	19	15
6						14	11	10	9
7							26	21	16
8								37	17
9									43
	Percentage similarity of species among the 10 sample plots inventoried in Ghana								
1	_	54	40	49	54	28	47	64	54
2		_	52	58	51	35	58	55	70
3			_	57	44	40	48	49	43
4				_	56	40	58	58	53
5					_	43	63	59	42
6						_	55	39	32
7							_	67	46
8								_	43
9									_

Mean ± standard error = 50 ± 2%

Table 16.4 Sorenson's species similarity per sample plot in Guinea

Plot	1	2	3	4	5	6	7	8	9	10
Comparative species richness among 10 sample plots inventoried in Guinea (underlined values in the diagonal indicate total richness of species in individual sample plots)										
1	41	15	24	22	21	18	17	17	12	15
2		28	14	12	14	9	11	10	8	8
3			39	27	23	20	18	17	13	18
4				33	18	19	21	17	14	17
5					35	16	15	11	7	11
6						27	17	13	10	15
7							28	13	8	13
8								24	15	14
9									20	15
10										25
Percentage similarity of species among the 10 sample plots inventoried in Guinea										
1	–	44	60	60	55	53	49	53	39	46
2		–	42	39	44	33	39	39	33	30
3			–	75	62	61	54	54	44	56
4				–	53	63	69	60	53	59
5					–	52	48	37	26	37
6						–	62	51	43	58
7							–	50	33	49
8								–	68	57
9									–	67
10										–
Mean ± standard error = 50 ± 2%										

Table 16.5 Rank order of comparable species in sample plots in Ghana and Guinea

No.	Ghana	Guinea
1	*Chromolaena odorata*	*Daniellia oliveri*
2	*Cissus* sp.	*Pericopsis laxiflora*
3	*Vertivera fulvibarbis*	*Andropogon gayanus*
4	*Vitellaria paradoxa*	*Detarium microcarpum*
5	*Borreria* sp.	*Parinari curatellifolia*
6	*Bridelia micrantha*	*Pterocarpus erinaceus*
7	*Grewia laxiodiscus*	*Prosopis africana*
8	*Annona senegalense*	*Sporobolus* sp.
9	*Anogeissus leiocarpus*	*Annona senegalense*
10	*Combretum* sp.	*Bridelia ferruginea*
11	*Daniella oliveri*	*Commelina* sp.
12	*Gardenia* sp.	*Danthoniopsis chevaleri*
13	*Indet*	*Gardenia ternifolia*
14	*Lannea nigricans*	*Lophira lanceolata*
15	*Terminalia albida*	*Quassia undulata*

Table 16.5 (cont.)

No.	Ghana	Guinea
16	*Vitex doniana*	*Swartzia madagascariensis*
17	*Aspilia africana*	*Afzelia africana*
18	*Aspilia* sp. (white flower)	*Crossopteryx febrifuga*
19	*Cussonia* sp.	*Dioscorea* sp.
20	*Desmodium giganticum*	*Parkia clappertoniana*
21	*Ficus sur*	*Terminalia albida*
22	*Indigofera* sp.	*Terminalia macroptera*
23	*Nauclea latifolia*	*Hymenocardia acida*
24	*Parkia clappertoniana*	*Piliostigma thonninguii*
25	*Setaria* sp.	*Vitellaria paradoxa*
26	*Strychnos spinosa*	*Xerroderis sthulmanii*
27	*Trichilia emertica*	*Bombax costatum*
28	*Triumffeta cordifolia*	*Cana indica*
29	*Uvaria chamae*	*Combretum nigricans*
30	*Anchomanes difformis*	*Monotes kerstingui*
31	*Bridelia ferruginea*	*Psorospermum febrifuga*
32	*Cassia siberiana*	*Raphiostylis benininsis*
33	*Lophira lanceolata*	*Smilax krausianus*
34	*Piliostigma* sp.	*Aframomum melanogueta*
35	*Tephrosia linearis*	*Albizzia zygia*
36	*Vernonia colorata*	*Anthonotha crassifolia*
37	*Allophylus* sp.	*Cajanus kerstingii*
38	*Byrsocarpus coccineus*	*Cochlospermum planchonnii*
39	*Clausena anisata*	*Costus spectabilis*
40	*Hyptis* sp.	*Cyperus* sp.
41	*Indet*	*Dissotis* sp.
42	*Justicia flavor*	*Landolphia heudelotii*
43	*Lannea macrocarpa*	*Lannea acida*
44	*Lantana* sp.	*Maytenus senegalensis*
45	*Tephrosia elegans*	*Pteleopsis suberosa*
46	*Tragia* sp.	*Uapaca somon*
47	*Abrus precatorius*	*Agave sisalana*
48	*Amorphophylus* sp.	*Amorphophalus* sp.
49	*Cassia hirsuta*	*Cantella asiatica*
50	*Cochlospermum* sp.	*Canthium venosum*
51	*Combretum* spp.	*Cassia siberiana*
52	*Commelina* sp.	*Cissus* sp.
53	*Desmodium* sp.	*Cyperus rotundus*
54	*Diospyros* sp. (broadleafed)	*Entanda africana*
55	*Fimbrystylis* sp.	*Erythrina senegalensis*
56	*Horrlahaena floribunda*	*Khaya senegalensis*
57	*Malacantha* sp.	*Lannea velutina*
58	*Mucuna* sp. (long red fruit)	*Leguminosea indet*
59	*Paspalum* sp.	*Strychnos spinosa*
60	*Roettboeillia cochichinensis*	*Albizzia adianthifolia*
61	*Sterculia sertigera*	*Aloe* sp.
62	*Tridax procumbens*	*Antidesma membranacea*
63	*Acacia nilotica*	*Asparagus pual*

Table 16.5 (cont.)

Nos.	Ghana	Guinea
64	*Afzelia africana*	*Cissus aralidoides*
65	*Amaranthus* sp.	*Cissus doeringii*
66	*Biophytum petersianum*	*Cola cordifolia*
67	*Cathormion altissimum*	*Combretum ghasalens*
68	*Clerodendron* sp.	*Cussonia* sp.
69	*Combretum* spp.	*Desmodium adscendens*
70	*Crinum* sp.	*Dioscorea preussi*
71	*Cyathula* sp.	*Ficus* sp.
72	*Dioscorea* sp.	*Indet* (*baa* in vernacular)
73	*Ehretia cymosa*	*Malnikara multinerus*
74	*Euphorbia* sp.	*Pavetta oblongifolia*
75	*Evovolus* sp.	*Psychotria rupifilis*
76	*Gmelina arborea*	*Saba senegalensis*
77	*Indet*-1	*Strychnos densiflora*
78	*Indet*-2	*Tephrosia* sp.
79	*Indet*-3	*Trichilia emertica*
80	*Indet*-4	*Vernonia* sp.
81	*Indet*-5	*Vitex doniana*
82	*Indet*-6	*Ximmenia americana*
83	*Indet*-7	
84	*Indet*-8	
85	*Indet*-9	
86	*Indigofera hirsute*	
87	*Khaya senegalensis*	
88	*Kyllinga* sp.	
89	*Lannea acida*	
90	*Lippia* sp.	
91	*Mitragyna enermis*	
92	*Panicum* sp.	
93	*Paullinia pinnata*	
94	*Rubiaceae indet*	
95	*Salacia* sp.	
96	*Securinega virosa*	
97	*Sida acuta*	
98	*Spathodea campanulata*	
99	*Sporobolus pyramidalis*	
100	*Tectona grandis*	
101	*Tephrosia* sp.	
102	*Terminalia glauscens*	
103	*Thonningia sanguina*	

Abundance diversity and Shannon-Wiener index

Generally, there was no significant difference in the abundance diversities of the species (Figure 16.5). This is evident in the proportional diversity by Shannon-Wiener index for the sampled plots. The indices were 2.88 for Ghana and 2.87 for Guinea.

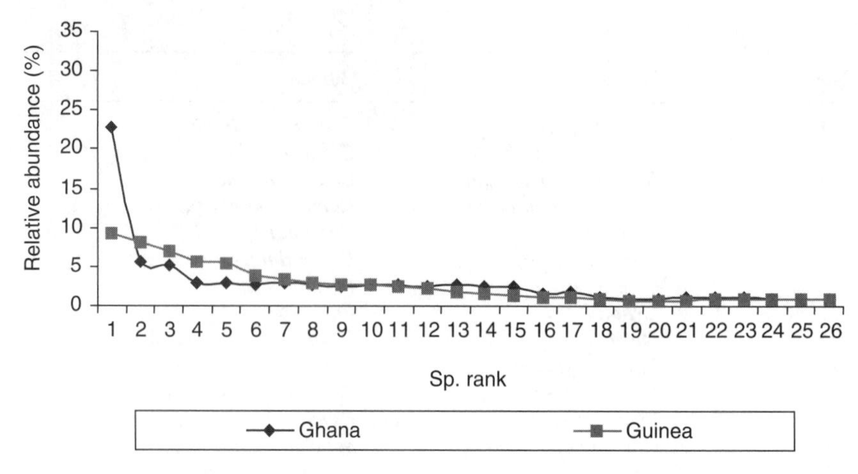

Figure 16.5 Abundance diversity of plant species in 0.4 ha study field

Girth-class distribution

Whereas *D. oliveri* was the most abundant species per plot encountered in Guinea, *A. leiocarpus* was the most abundant in Ghana.

The girth-class distribution of woody species in the plots inventoried in the two countries is presented in Figure 16.6. Generally regeneration of woody species is high in the plots studied, particularly in Guinea (Kouroussa-Moussaya). Similarly, there is an appreciably higher representation of pole-size wood species (50–100 cm) in Guinea.

In general, the representation of girth classes that are between 150 and 200 cm is twice to thrice as much in the plots studied in Guinea than in Ghana. For example, except for Kouroussa-Moussaya, larger stems (200–250 cm) are less represented and even absent altogether in Asantekwa, whereas in Binguri, Bognayili, and Dugu-Song (all in Ghana), this girth class is represented only in small numbers. Again, the 250–300 cm girth-class wood species is also better represented in the plots at Kouroussa-Moussaya in Guinea than in the plots sampled in Ghana. Therefore, representation of much larger stems (>300 cm) in plots studied in Ghana is estimated to be less (i.e. 33–50 per cent) than what was observed in plots in Guinea.

Height-class distribution

The protected forest at Binguri showed twice as many regenerating tall species than at Kouroussa-Moussaya (Figure 16.6). A progressive increase in pole-size numbers at Dugu-Song and in other plots in Ghana is observed (Figure 16.7). By contrast, there appears to be a successive reduction and loss in numbers of

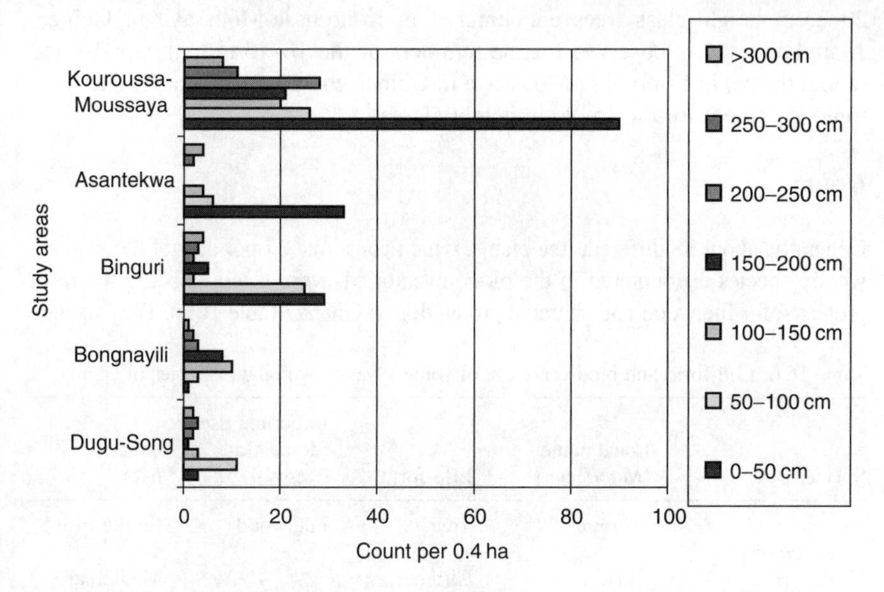

Figure 16.6 Comparative girth-class distribution of woody trees

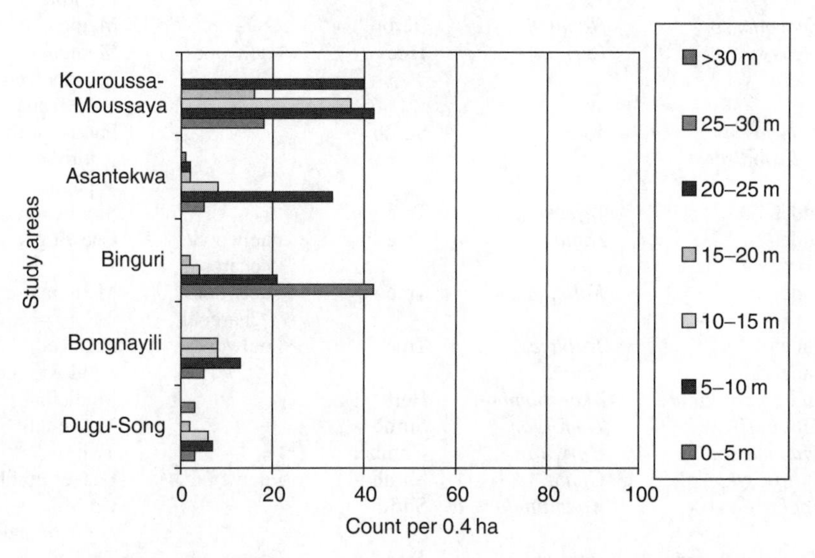

Figure 16.7 Comparative height-class distribution of woody trees

20–25 m height-class trees encountered in Kouroussa-Moussaya in Guinea. A similar trend is observed for the numbers of the 10–15 m high wood-stand (about thrice) in Kouroussa-Moussaya in Guinea compared to Ghana. The same trend is evident for the 15–20 m height classes.

Utility

Generally, about 38 different tree plants, which constitute 37 per cent of the savanna woody species encountered in the plots inventoried, were listed as being the most preferred for fuelwood and charcoal production in Ghana (Table 16.6). They include

Table 16.6 Life-form and biodiverse use of some savanna woodland species in Ghana

Species	Local name (Moo/Dega)	Life-form	Important use (domestic energy)	Other uses
Annona senegalense	*Mormor*	Tree	Fuelwood	Edible fruit
Cissus sp.	*Tsagba*	Climber		Medicinal
Cochlospermum sp.	*Kpankpanhuru*	Shrub		Roots used as spice
Cussonia sp.	*Balasiewu*	Tree		Craft
Desmodium sp.		Herb		Cough medicine
Dioscorea sp.		Stem tuber		Not edible
Diospyros sp.	*Kanu*	Tree	Fuelwood	Timber, fruits edible
Evovolus sp.	*Hanabel*	Herb		Medicinal
Ficus sur	*Purr*	Tree	Fuelwood	Medicinal, birds feed on fruits
Grewia laxiodiscus	*Yole*	Shrub		Food dye for pito/local paint
Indet-1	*Yeyewu*	Tree		Stool carving
Indet-2	*Hinla*	Tree	Fuelwood/ charcoal	Chewing-stick
Indet-3	*Kologna*	Tree	Fuelwood/ charcoal	Medicinal
Indet-4	*Torbuge*	Tree	Fuelwood	Fruits edible
Indet-5	*Tibanu*			Food wrapper
Indigofera hirsuta	*Akarabombo*	Herb		Medicinal
Lippia sp.	*Timforgor*	Shrub		Medicinal
Mucuna sp.	*Horwuro*	Climber		Twine
Piliostigma sp.	*Gbate*	Shrub	Fuelwood	Leaves edible
Tephrosia sp.	*Kpanatinda*	Shrub		Craft/ medicinal
Trichilia emertica	*Borgor*	Tree	Fuelwood	Medicinal
Vetivera fulvibarbis	*Kuukuma*	Grass		Fodder

the following species: *A. leiocarpus*, *A. senegalensis*, *Diospyros*, *Piliostigma*, *Terminalia* sp., *Trichilia* sp., *Ficus* sp., *L. lanceolata*, *Sclerocarya birrea*, *Vitex* sp., *L. acida*, *Combretum* sp., *Detarium* sp., and *Balanites* sp. There is also charcoal production in both countries. However, it was observed to be much higher in Ghana.

Other uses of savanna woodland resources include medicinal ones, roots used as spices in cooking, timber for construction, wood for carving, chewing-sticks, edible leaves and fruits, and oil extraction. In addition, a few selected trees are left on the farmlands as agroforestry species (Table 16.6) in combination with crops.

Discussion

Species richness

Species richness is higher in Asantekwa in Ghana than at Kouroussa-Moussaya in Guinea. This may partially be attributable to the extensive gaps or breaks in the canopy of the forest of the transition zone or the savanna woodland in Asantekwa (Ghana), which result from the harvesting of wood either directly for fuelwood or for the production of charcoal to meet household energy requirements. The wood harvesting, in turn, leads to the influx of opportunistic species, particularly those whose seeds are wind dispersed such as *C. odorata*, *Aspilia* sp., and *A. africana*. Significantly, these species were not found in sample plots in Guinea. It is also conceivable that the observed differences in species richness are due to natural variations in climatic conditions in the two study areas.

General rank order of species

The observed higher ranking of *D. oliveri* in Guinea may be attributed to the differences in management practices of the vegetation resources in the two countries. In Kouroussa-Moussaya in Guinea, the area studied was a reserve under protection where early burning is practised, whereas the plots studied in Ghana did not appear to be under any form of protection, and fire was not used as a deliberate ecological management tool. The use of fire is not controlled in Ghana. For reasons that still remain ill-understood, early burning in the plots in Guinea did not seem to affect regeneration and pole-size classes of woody species.

It is also likely that the relatively low rank of *L. lanceolata* in the plots studied in Ghana may be due to its high preference as a timber species as well as for mine props, railway sleepers, and for charcoal production. In the same way, the rather high ranking of *C. odorata* in the plots in Ghana may be indicative of the extensive and rather careless opening up of the canopy with a concomitant invasion of opportunistic species.

The rather high rank of *V. paradoxa* in Ghana could be due to the fact that the species has economic value for the production of shea butter for commercial purposes, and is therefore left protected in place so that its fruits may be harvested for sale.

The absence of pole-size classes of the different woody species encountered in the sample plots at Kouroussa-Moussaya may be due to the use of fire in the early-burning vegetation management practice.

Abundance diversity and Shannon-Wiener index

The observed small numbers of a large number of species in the plots studied in both countries are characteristic of the natural heterogeneous ecosystem globally. However, the observed abundance diversity of *D. oliveri* in Guinea, which is three times higher than in Ghana (0.34 and 0.11, respectively), can be attributed to the protection of the study area as a reserve in Guinea. In the same way, the several-fold higher abundance diversity for *V. paradoxa* in the plots in Ghana can be attributed to the protection accorded to this species due to its economic value for shea butter production as noted above. The observed similarity in the values of abundance diversity of *P. clappertoniana* for Guinea and Ghana may be due to the similarity of protection of the species as an important food condiment.

Girth, height-class distribution, and utility

The observed high representation of individuals of each of the larger girth (200–250 cm) and height classes (20–25 m) in the plots in Guinea appears to be related to the type of management practice, which, in Guinea, involves the protection of the study area as a reserve, unlike in Ghana where no such protection occurs. Consequently, the high demand for woody species to supply household energy requirements either as fuelwood or charcoal has led to the denuded state of the vegetation in the study areas in Ghana.

Aside from use as fuelwood, most of the species encountered in the plots studied in the two countries have value as medicine, dyestuff, or food colour, besides being commercial sources of edible fruits.

While fencing of the household was not a common feature of the settlements in the vicinity of the plots studied in Ghana, it was a major feature in such settlements in Guinea.

In Ghana, the high population density and a concomitant high demand for fuelwood and charcoal as a domestic energy source may be contributing to deforestation and adverse changes in the microclimate.

It is arguable that the savanna woodland can turn into forest when protected (Eyre, 1968; Bourlière and Hadley, 1983). However, vast areas of this woodland, particularly in the case of Ghana, are becoming thinned and impoverished as a consequence of intense harvesting for domestic energy needs, conversion into agricultural fields, and, more importantly, unrestrained burning through wild fires in the dry season.

Where there is protection of savanna woodland, as in Kouroussa-Moussaya in Guinea, and Binguri and Jaagbo in Ghana, an increase in the richness of species, regenerating saplings, and number of pole-size individuals of different species is observed.

On the whole, therefore, the size-class structure of the savanna woodland in Guinea appears to be more stable and typical of a natural "forest" or savanna woodland than the corresponding areas studied in Ghana. The differences observed above may be due to differential performance of the same species under different abiotic and other ecological conditions.

In Ghana, absence of pole-size individuals coupled with continuous harvest of woody species in the plots in Asantekwa, Bognayili, and Dugu-Song will have great implications for recruitment into the structure of the forest in future, in terms of availability of these woody species, if alternative measures of management are not developed and implemented.

Conclusion

The floral compositions of the savanna woodland in the Republics of Ghana and Guinea are similar. The entire sample areas lie in the forest-savanna transition zone of West Africa and cover virtually the same latitude. There is no significant difference in the abundance diversities of the species identified in the plots studied in the two countries. However, the frequency distribution pattern of species shows slightly different forms due to differences in management and protection practices of these woodlands.

The girth-and-height size class differences of species observed may also be due to differential performance of the same species under different abiotic, anthropogenic, and other ecological conditions. For example, in Ghana, absence of pole-size individuals in the plots in Asantekwa, Bongnayili, and Dugu-Song, coupled with continuous harvesting of woody species, has affected cohort recruitment for the establishment of any meaningful forest structure (Swaine and Hall, 1986). On the contrary, in Guinea a vast area of the savanna woodland is under protection as a forest reserve. Therefore, the cohort classes of woody species in sample plots at Kouroussa-Moussaya are closely related and thus have a continuous canopy compared to Ghana.

The high population with its attendant high demands for fuelwood and charcoal as a domestic energy source may be contributing to the denudation of the savanna woodland in Ghana. Whereas settlement and household fencing using large stem cuttings is a common practice in Guinea, this practice does not feature prominently in the settlements of the study area of the savanna woodland in

Ghana. The intensity of the practice in Guinea is limited to the relatively less populated areas of the province around Kouroussa-Moussaya.

If collaborative management efforts, particularly workable policy measures that cut across regional frontiers of West Africa, are not radically promoted, then the resource potential and the ecology of the transition zone stands a great chance of being threatened. The rapid population increases linked to the high demands for vegetation resources and land use in the subregion calls for prompt action. For example, considerations of win-win scenarios in the different neighbouring countries needs to be reviewed in the context of sustainable vegetation resource utilization within the Economic Community of West Africa (ECOWAS). Alternate domestic energy sources need to be explored and improved. Cross-border research towards effective and efficient management must be promoted. The good lessons learnt must be replicated. By so doing, the problem of land degradation in the savanna woodland of the subregion could be reversed.

REFERENCES

Bourlière, F. and P. Hadley, "Present-day savannas: An overview", in F. Bourlière, ed., *Ecosystems of the World 13*, *Tropical Savannas*, Amsterdam: Elsevier, 1983, pp. 1–17.

Coffey, K., *PLEC Agrodiversity Database Manual*, Report for PLEC, United Nations University, 2000, available from www.unu.edu//env/plec.

Eyre, S. R. *Vegetation and Soils*, in A. T. Bâ, J. E. Madsen, and B. Sombou, eds, *Atelier Sur Flore, Végétation et Biodiversité au Sahel*, AAU Reports 39, Bath: Pitman Press, 1968.

Swaine, M. D. and J. B. Hall, "Forest structure and dynamics", in G. W. Lawson, ed., *Plant Ecology in West Africa: Systems and Processes*, Chichester: John Wiley & Sons, 1986, pp. 47–93.

Thies, E., *Principaux ligneux (agro-) forestiers de la Guinée. Zone de transition: Guinée-Bissau, Guinée, Cote d'Ivoire, Ghana, Togo, Benin, Nigeria, Camarou*, Rossdorf: TZ-Verl. – Gess, 1995.

Zarin, D. J., H. Guo, and L. Enu-Kwesi, "Methods for assessment of plant species diversity in complex agricultural landscapes: Guidelines for data collection and analysis from the PLEC Biodiversity Advisory Group (BAG)", *PLEC News and Views*, No. 13, 1999, pp. 3–16.

17

Agrodiversity within and without conserved forests for enhancing rural livelihoods

Essie T. Blay, Benjamin D. Ofori, John Heloo, Joachim B. Ofori, and Emmanuel Nartey

Background

In order to encourage *in-situ* conservation of biodiversity and retention of some forest cover in agricultural areas, it is imperative to ensure that the farmers derive some economic livelihood from conserved forests. A key strategy in this regard is integration of commercial activities into the conserved forest. Among the major candidate ventures in Ghana are:

- apiculture or keeping of bees for honey and related products
- snail rearing
- grasscutter production
- yam cultivation within the conserved forest, or outside it, as may be the case.

This chapter discusses these activities on the basis of PLEC research work in Ghana.

Apiculture in conserved forests

Income generation from honey is an age-old practice in rural areas of Ghana. Various traditional methods of beekeeping exist countrywide. Honey-hunters traditionally comb beehives of wild bees in branches of trees in the forest, dead tree trunks, caves, eaves of houses, etc. for honey that is used in the home or sold for direct consumption or medicinal purposes.

The recent spate of health consciousness in Ghana has created a big market for honey as a sugar substitute. However, typically, the honey from wild sources is of poor quality because of crude methods of harvesting, extraction, and handling. Its supply is also uncertain.

To ensure a more reliable source of honey, some farmers in rural areas have adopted beekeeping on their farmlands. This is rapidly gaining ground in the forest and savanna woodland zones of Ghana as a means of generating income from conserved forests.

The African honeybee, which is cultured in Ghana, belongs to the species *Apis mellifera adansonii*. It is smaller than the temperate races of *A. mellifera*, but has a longer proboscis and is a more flexible forager. This makes it potentially more efficient at honey production than its temperate counterparts (Adjaloo and Yeboah-Gyan, 1991). However, this potential is yet to be tapped because apiculture based on this strain is not well researched and developed (Oppong, 1991). Furthermore, the African honeybee is temperamental, ferocious, and prone to absconding, which makes it more difficult to manage (Adjare, 1991).

The governments of Africa generally and Ghana in particular recognize the economic value of honey and have identified apiculture as a relatively low-input income-generation system for improving rural livelihoods. Ghana's Ministry of Agriculture made the maiden attempt at developing beekeeping in Ghana in the early 1960s with the introduction of several colonies of temperate honeybee races. However, none of the colonies survived. Therefore beekeeping was shelved until 1978 when, with the help of some literature from Kenya, Dr J. W. Palmer, the then director of the Technology Development Centre of Ghana, took steps to revive it. Since the late 1980s beekeeping has steadily gained popularity, and more sophisticated production systems have also been introduced.

Beekeeping practices in the rural areas

Traditional systems for beekeeping include the use of large earthenware pots (Plate 13). The hollow pot, which houses the honeycombs, is provided with a tight cover and an entrance hole for the bees. The baited or colonized pot is placed on a stand under forest shade, preferably close to a source of water. In areas removed from natural water sources, the beekeeper provides a constant supply of clean potable water in a bowl.

Other indigenous apicultural systems practised in rural areas in Ghana include the use of dugout logs, specially woven baskets, kerosene tins, and mats of thatched corn stalks formed into a cylinder and provided with a tight-fitting lid bearing an entrance hole for the bees. These structures are baited with bee wax, palm wine, perfume, cow dung, lemon grass, or other suitable substances, and exposed in tree boughs or on supports in the forest for colonization by wild bees. Under suitable environmental conditions, colonization occurs with ease. Where colonization is difficult, a specially designed gourd or small earthenware pot is baited and used to trap bees, which are then transferred into the beehive.

In recent years improvement in apiculture in tropical Africa has led to increasing adoption of the Kenyan top-bar beehive constructed from hardwood (Plate 4). Other types of improved beehives in use are the Langstroth hive and the Tanzanian transitional long hive, but these are less common.

Rural beekeepers have extensively adopted the use of the Kenyan top-bar hives. These are kept in the agroforestry home gardens, under trees in the farms, on fallow lands, or in conserved forests, sometimes along with the traditional beehives.

Feeding

The natural fauna in the farms and around farmsteads form the base of forage for the honeybees. The African honeybee is a ubiquitous forager. Its long proboscis enables it to forage from flowers with short or long corollas (Adjaloo and Yeboah-Gyan, 1991). A wide range of melliferous plant species found in the forest, savanna, and forest-savanna transitional areas offer suitable sources of nectar and pollen. Common forage plant species include agricultural crops like oil-palm, mango, orange, lime and other citrus plants, pineapple, guava, cashew, cassava, solanaceous, and cucurbit vegetables. Other important non-cultivated or semi-cultivated species such as the shea tree, silk cotton, mahogany, lantana, and a large number of flowering trees, shrubs, and grasses also constitute an important component of the forage crops. The wide diversity of non-crop and crop plants on the peasant holdings provide year-round sources of feed for the honeybees. The farmers also introduce species that are known to support beekeeping into the farmlands, and they take steps to protect the natural vegetation to create an enabling environment for the bees, enhance their feeding, and prevent them from absconding. The African bee, unlike the European bee, is not known to die in its hive from starvation. They abscond from the hives for a better environment when threatened by starvation. This makes it imperative to provide optimum management practices, which will curb swarming. Little, if any, supplementary feeding with sugar syrup is practised.

Harvesting

The cool period of night is regarded as the best time for honey harvesting. During harvesting, honey-hunters smoke the hives of wild bees to drive them out. The combs are subsequently collected and squeezed to release the honey, which is stored in beer bottles or gallon-size plastic bottles for sale. Many rural operators of the Kenyan top-bar beehive system also collect and squeeze the combs to release honey. Honey obtained this way has poor quality and hence low market value. The system adopted by commercial beekeepers involves the use of special smokers to immobilize the bees, after which the honeycombs are removed from the beehives and centrifugal extractors are used to release the honey.

Harvesting is done several times a year except during the rainy season. The major harvesting periods in southern Ghana are February, March, April, November, and December. The farmers receive extension training and support for proper harvesting techniques from extension officers of the Ministry of Agriculture, some NGOs, and from expert farmers in PLEC associations of farmers.

Disease and pest management

The African honeybee is an efficient housekeeper. The hive and the entrance are continuously cleaned and waxed by the bees. This considerably reduces disease incidence. However, a number of pests have been reported. These include wax moths, wasps and hornets, ants, lizards, and mites. Disease control is achieved by maintaining a sanitary environment around the hives.

Benefits derived from beekeeping

Beekeeping does not only offer a tangible source of income for the rural farmer, it also plays a fundamental role in conservation of natural vegetation. The benefits of beekeeping include the following:

- increased pollination of flowering plants, increased seed set and yield in flowering agricultural plants, and regeneration of forests and other vegetation from seeds
- higher crop yields resulting from enhanced pollination by honeybees
- stimulation of interest in conservation of forests and other vegetation that form the base of the industry
- creation of job opportunities for beekeepers who provide supporting services
- improvement in farmer livelihoods through enhanced income, food security, nutrition, and health
- income from the sale of honey, the primary product
- income from the sale of medicinally useful propolis, a sticky, gummy, resinous substance that the bees gather from barks and buds of plants during foraging for use in mending the beehive and sealing cracks in it
- income from beeswax, which is used for manufacturing candles
- energy derived by human beings from consumption of honey
- traditional medicinal uses of honey, notably in the treatment of common coughs
- potential of becoming an important source of foreign exchange.

Benefits associated with PLEC-supported beekeeping by smallholder farmers in Sekesua-Osonson, a PLEC demonstration site in southern Ghana, are discussed by Gyasi and Nartey (2003).

Constraints

The major constraints on development of beekeeping in Ghana include the following:

- lack of technical competence
- limited information on the African honeybee
- absconding of bees
- phobia about bees
- limited capital for acquisition of the proper gear for harvesting and processing the honey
- bushfire, which destroys the habitat of the bees
- use of agrochemicals that are inimical to bees
- marketing problems emanating from poor quality of the honey produced and ignorance of marketing channels.

Raising snails in conserved forests

Background

The giant African snail is an important source of animal protein in many parts of Ghana. The black strain (*Achatina achatina*), locally known as *nwapa*, is the largest. It occurs commonly in the forest areas, while the lighter-coloured species *mpobri* or *otabraa* (*A. marginata*) is more predominant in the savanna areas. The snail has a high percentage of good-quality protein (69 per cent dry weight) and abounds in minerals such as potassium and phosphorus, and in vitamins C and B complex. However, certain tribes, notably the Ewes, others in northern Ghana, and certain religious groups, do not eat the snail.

Traditionally rural folk scout freely in the forests and farmlands to collect snails during the rainy season for sale and domestic consumption. Snail collection is usually undertaken during the night or in the early hours of the morning.

Snail, both live and processed, attract very high prices locally and abroad, where they are increasingly exported. This trend has generated a lot of interest in snail rearing in both urban and rural areas. In the rural areas snail fattening as well as snail rearing are practised in the homestead or on farmlands. This has the promise of engendering interest in natural vegetation conservation to provide a ready source of breeding material and a suitable environment for rearing the snails. Farmers also tend to protect tree species that are known to attract snails and, sometimes, they intentionally introduce these species into their farms. Snail rearing is an easy, reliable, and low-input source of supplementary income from conserved forests. For these reasons conservationists, wildlife experts, rural developers, government agencies, NGOs, and other organizations such as PLEC are promoting biodiversity conservation in the rural areas by supporting snail farming within and outside conserved forests.

Sources of snails for rearing

In the dry season snails bury themselves into loose soil or decaying vegetation, enclose their shell in a thin, white membrane, the epiphragm, and hibernate. They normally emerge from hibernation during the rainy season between the months of March and October when conditions are moist and conducive for their growth. During this period snail producers collect their stock from the wild in the undergrowth of plantains, banana, coffee, and cocoa farms and in forest undergrowth. Some snail collectors report that debarked branches of *onwe* are stuck in the ground to attract the snails, thereby facilitating their collection. In the forest, snails are also found under *Ficus* species on whose fruits they feed. Alternately, snail farmers purchase their stock from the market or from other producers anytime during the year.

Housing

For housing the snails, two systems are used: an indoor system and pasture/outdoor system. In the modern indoor system, snails are cultured in plastic containers, which are arranged on wooden shelves or galvanized iron pipes. They may also be raised in wooden boxes. Container sizes are variable. A 60 × 30 cm container can house about 100 snails weighing about 50 g each. The housing units are provided with entrance for tending the snails, and small spaces are allowed to let in dim, diffused light since the snails are sensitive to high light intensity. All open spaces around the containers are covered with wire netting to prevent the snails from escaping and to exclude predators. The bases of the containers are covered with moist sandy, loamy soil and some leaf litter to simulate the natural habitat and to allow them to burrow and prepare their laying pit.

After laying, the eggs are collected and kept in breeding boxes under shade or in a room and provided with a warm, humid environment to facilitate hatching. The eggs hatch after 15–40 days and the newly hatched snails are collected and raised in another box until they are big enough to be transferred into the fattening pens.

Farmers in the rural areas can easily adopt the indoor system since the rural housewife is used to maintaining a few snails in containers close to the kitchen for family use on a short-term basis. Such a project may be sited in the backyard and the kitchen waste may be conveniently used as supplementary feed. Containers that are easily available in the villages such as large clay pots and strong woven baskets with appropriate lids may be used to house the snails.

The modern outdoor/paddock system for raising snails involves clearing the site and marking out the area to be fenced. The snails are then confined in paddocks constructed in the cleared area by driving wooden posts into the ground and covering them with plastic sheets.

Farmers in the rural areas practise various forms of the outdoor system of snail raising, which are discussed as follows. In conserved forests, enclosures are made under the shade of single large trees or clumps of trees in moist areas, which are the natural habitats of the snails. The enclosure is made by driving wooden posts into the ground to create a large, circular paddock around the tree or clump of trees. Where clumps of trees are used, the trees at the periphery of the clump may serve as the posts. The posts are covered with plastic sheets or nylon netting, the base of which is buried deep into the soil to prevent the snails from crawling under. The fallen leaves from the tree create a litter bed, which forms a suitable micro environment for the snails. Crops such as cocoyams, banana and plantain, and other non-crop plants which form a ready source of feed for the snails are allowed to grow freely within the fenced-off area. A source of water supply is provided and crop stubble and harvest leftovers are served as supplementary feed. This system closely mimics the natural environment of the snail and allows for the high stocking rate. Such a system is in practice by Bossman Kwapong in Gyamfiase, Adenya, a PLEC demonstration site in southern Ghana. A second practice or system involves the use of pits. In this, a rectangular pit, 1 m wide, 1 m deep, and of any convenient length is dug under shade in the home garden. The base of the pit is gently sloped to facilitate drainage during the rainy season. A layer of brick or a concrete wall is provided above ground and fitted with a cover to keep off excessive rain and predators. Alternately, a short nylon net cage is fitted on top of the pit to protect the snails. The base of the pit is covered with loose soil and leaf litter. Many variations of this system are practised, including the use of different sizes of circular pits and different systems of confining the snails to the pits. The eggs are removed and hatched in a different pit and the newly hatched are transferred into another pit until they are large enough to be sent into the fattening pit. The pit system promises to provide a better security for the snails than the free-range enclosure system. A possible disadvantage of this system is flooding under heavy rains. Siting the pit in well-drained areas, erecting a high enough wall around it, and providing appropriate roofing may prevent this problem.

Feeding

There is no problem with the feeding of the snails on the farm. The giant African snail is essentially vegetarian and browses on a wide range of leaves. It also feeds on various fruits. Almost all edible tropical drupes are acceptable to them. They also feed on all edible vegetables, vegetable clippings, leafy vegetables, cotton leaves, banana and plantain leaves, castor oil plant, cassava peel, and yam tubers and leaves. In addition to this, compounded feeds are available on the market for enhancing the fattening process where very large-scale production is anticipated. Farmers protect or introduce some indigenous tree species such as the *Ficus* sp. that provide a good source of feed for the snails. The snails should be provided with a regular, easily accessible source of clean water.

Pests and disease management

The major pests are snakes, frogs, soldier ants, and mites. The housing should be designed and its environment managed in such a manner as to exclude these pests. Left-over feed should be promptly removed from the pen and the pen kept clean and moist to avoid disease problems.

Grasscutter production

Background

Production of the grasscutter (*Thryonomis swinderianus*) is a very viable income-generation venture that is compatible with small-scale peasant farming systems.

Grasscutter meat, a major source of animal protein in the landlocked rural areas in Ghana, is regarded as a delicacy countrywide and is becoming important on the small-scale food export market. The grasscutter is about the size of a rabbit. It is a member of the rodent family. Two strains occur in Ghana. One is adapted to the forest areas and the other occurs in the savanna. They are found in the wild associated with areas abounding in wild graminaceous plants, or on farmlands where yams, cassava, sweet potato, cereal grains, etc. are cultivated. Traditionally grasscutter meat is obtained by trapping, or from hunting involving free use of fire to scare the colonies. The latter has been a major source of bushfires with attendant disruption of the ecosystem.

Grasscutter raising is promoted by the government of Ghana and certain NGOs, notably the German Technical Cooperation (GTZ), the Catholic Church, the Heifer Project International (HPI), and private individuals with the aim of turning grasscutter hunting around and making controlled production an important source of income for the rural population. This is seen as a major step in stemming the rampant bushfires that occur during the peak of the grasscutter-hunting activities in the dry season. Several factors make grasscutter production an excellent project for supplementing rural livelihood and promoting conservation of natural vegetation. The low input requirement for grasscutter raising makes it a very suitable income-generation activity for the small-scale farmer. Low-cost housing can easily be constructed from materials available on the farm. Sources of feed also abound in the farmstead and the produce has very high demand.

Breeding stock

The two main sources of breeding stocks in Ghana are:

- animals caught in the wild, using various traps such as metal cages, and concealed dugout holes in grass-covered areas with or without ammonia (urine) fumes as an attractant
- the progeny of already domesticated animals from the few approved grasscutter-raising farms in Ghana, and some breeding animals imported from Benin, where grasscutter production is well developed (Annan and Weidinger, 2001).

The use of breeding animals caught in the wild has proved less efficient because the catches as well as the quality of animal(s) are not guaranteed. The captured animals are traumatized, thereby increasing the mortality rate during the first few days after capture. Again, farmers experience a lot of difficulty in trapping the animals with metal cages because the grasscutters recognize the cages as traps and avoid them. The animals will only venture into very old worn-out cages, which appear abandoned. Furthermore, in the first few days of captivity the animals caught from the wild tend to throw themselves against the restraining walls of their pens in a bid to bolt, which leads to injuries and a high initial mortality rate.

Where breeding animals are obtained from grasscutter farms, there is the added advantage of the opportunity to select the breeding stock.

Guidelines used for selecting breeding stock include characteristics like temperament, fertility, litter size, growth rate, and vitality. Selected breeding stock must be free of any abnormalities and deformities.

Sexing and sex ratio

Grasscutter producers recommend a breeding male:female ratio of 1:4–9. To distinguish between the male and female the following characteristics are used:

- head shape and size: the mature male has a bigger head than the female
- ano-genital distance: the distance from the anus to the clitoris in the female is considerably shorter (7 mm) than the distance from the anus to the penis in the male (20–30 mm)
- red/yellow patch occurs around the genital area of the male.

Handling

For sexing or other purposes, grasscutters are handled with a great deal of caution to avoid attack with their powerful incisors, or with the claws on their hind legs. The recommended methods of handling are holding the tail, the neck (scruff), the rump (waist), and the backbone.

Housing

Grasscutters are housed in hutches (cages) constructed with wooden frames using hard wood. This is lined with a hard wire mesh that cannot be easily chewed with the grasscutter's powerful teeth. The hutches are placed individually, arranged in single rows or put together in tiers.

An alternative involves a low-cost housing, whereby mud bricks are used to construct a double-wall box about 1.2–1.4 m under a hut. The box is provided with a wooden lattice cover that may be partitioned to house one or two females, or up to nine females per compartment. Each cubicle is provided with an entrance at the top.

Feed and feeding

Grasscutters do well on very succulent vegetation. They possess a rather narrow throat and hence can only swallow small pieces of feed at a time. For this reason there is a very high percentage (70 per cent) of feed wastage. Under domestication, grasscutters are fed on fodder grasses like elephant grass (*Pennisetum purpureum*) and Guinea grass (*Panicum maximum*) on which they normally feed in the wild together with other fodder species including *leucaena* (*Leucena leucocephala*), *gliricidia* (*Gliricidia sepium*), and *stylosanthes* (*Stylosanthes guinensis*), pineapples, tubers of yam (*Dioscorea* sp.), cassava (*Manihot esculenta*), sweet potato (*Ipomea batatas*), cereal grains, maize stovel, and husk. They may also be fed on left-over meals and industrial waste. In addition, specially compounded feed may be used as a supplement (Table 17.1). When fed on succulent feeds, the grasscutter requires little water. However, a ready supply of good clean water is provided, especially where dry feed supplement is served.

Reproduction and disease control

The females reach sexual maturity at about six months. Males are ready to serve as breeding animals at seven months at a mating ratio of four females to a male. Animals on heat are sent into the male pen for crossing. After successful mating

Table 17.1 Concentrate feed supplement for grasscutters

Feed ingredient	Quantity (kg)
Maize grain	10.0
Wheat bran	15.0
Oyster shell	2.0
Salt	0.5

Source: Annan, Duku, and Sulemana (2002)

the gestation period lasts for 140 to 160 days. A litter of five to nine is normally produced. The babies are breastfed for about two weeks, after which they are ready to go on grass feed. The young grasscutters are weaned after six to eight weeks, after which the mother is rested for two to three weeks and crossed again when on heat. The interval between litters is roughly seven months. The females remain reproductively active for a minimum of five years.

Little information is available on the diseases encountered on peasant farms. However, with proper sanitation and isolation practices, disease attack is not a major problem.

Marketing

The animals are ready for slaughter from 2 kg body weight and above. The females attain about 2.5 kg and the male around 4 kg at 18 months. The grasscutter meat is marketed whole. Opportunities exist for marketing live animals when inexpensive methods of restraining them during transportation are developed.

Cultivation of yams in conserved forests

Yam is an important staple food crop in Ghana and elsewhere in West Africa. Its diversity and ways of managing it for food security are discussed with special reference to southern and northern Ghana in Chapters 7 and 8.

Yam species such as the bush yams (*D. prahensilis*), the bitter yam (*D. dumeturum*), and the aerial yams (*D. bulbifera*) may be integrated into forest agriculture by growing them along with forest trees or compatible tree crops like cocoa, as a reliable year-round source of food and for supplementary income generation. The adaptation of the bush yams to forest conditions offers an opportunity to incorporate them into a fallowed forest alongside other income-generating activities such as snail raising and beekeeping. This will ensure that farmers continue to derive some income from their lands during the fallow period and, thus, give them an incentive to lengthen the dwindling fallow periods.

The yields from the bush yams and water yams are very high and they constitute a significant source of income to the farmers. However, several varieties have very short storage life and losses are high. If the processing aspect is tackled to ensure a good market outlet, yam is a crop whose production can be increased by the farmers with little or no disruption to the environment.

In some parts of Ghana yams are grown in monoculture on clean cultivated fields and staked with bamboos, branches of trees, or stems of young trees that are harvested from the wild. This system is inimical to both soil and biodiversity conservation. The use of specially conserved trees, notably *Newbouldia* sp., for staking yams in traditional agroforestry systems, especially in Sekesua-Osonson PLEC demonstration site and other areas occupied by migrant Krobo farmers, appears to be a sound practice that serves the purpose of supporting food production while maintaining biodiversity and providing soil cover.

Conclusion

Agrodiversity, including diversification by beekeeping and by raising of snails, the grasscutter, and yams, especially if carried out on a commercial scale within conserved forests, holds considerable prospects of enhancing rural livelihoods while conserving the natural flora. It therefore deserves encouragement in the rural development strategy.

REFERENCES

Adjaloo, M. K. and K. Yeboah-Gyan, "The foraging strategies of the African honeybee *Apis mellifera adansonii* in the humid forest", *PLEC News and Views*, No. 20, 1991, pp. 36–43.

Adjare, S. O., "Apiculture development in West Africa: Ghana's experience", in N. Bradbear, ed., *Proceedings of a Scientific Workshop for West African Bee Researchers, "The First West African Bee Research Seminar"*, 1991.

Annan, P. and R. Weidinger, *Keeping Grasscutters for Benefit*, MOFA-Ghana, Technical Paper, 2001.

Annan, P., G. Duku, and O. Sulemana, *Grasscutter Rearing Manual*, Sedentary Farming Systems Project, MOFA, Brong Ahafo Region, 2002, p. 8.

Gyasi, E. A. and E. Nartey, "Adding value to forest conservation by bee-keeping at Sekesua-Osonson demonstration site in Ghana", *PLEC News and Views*, News Series 1, March 2003, pp. 12–14, available from http//c3.unu.edu/plec/; also, http://rspas.anu.edu.au/anthropology/plec.html.

Oppong, S., "Research priorities for beekeeping in West Africa", in *Proceedings of the First West African Bee Research Seminar*, 1991, pp. 58–60.

Part III

Social dimensions of resource management

18

Aspects of resource tenure that conserve biodiversity: The case of southern and northern Ghana

Edwin A. Gyasi and William J. Asante

Introduction

Resource access or tenure refers to the terms, arrangements, or rules and regulations governing control of or access to natural biophysical attributes regarded as valuable by society. The character or system of resource access is widely assumed to be an important determinant of the integrity of biophysical resources because the incentive to conserve them is, to a large extent, inherent in it (Gyasi, 1994).

This chapter discusses resource tenure in relation to conservation of biodiversity. It does so mainly on the basis of information from studies carried out under PLEC between the years 1993 and 2001 in the demonstration sites in Ghana:

- Gyamfiase-Adenya, Sekesua-Osonson, and Amanase-Whanabenya, which are located in the southern sector of the forest-savanna transition zone
- Bongnayili-Dugu-Song and Nyorigu-Binguri-Gonre in the northern savanna zone (Map A; see also Map C).

Overview of the tenurial arrangements

In the five selected demonstration/study sites, the following are identified as the principal types of resource tenure:

- land tenure, i.e. tenure with respect to "terra firma", the solid earth, which has special significance because it is the fundamental natural resource for agriculture, the most important economic activity that engages over 70 per cent of those employed

- tree tenure, i.e. tenure with respect to economic trees, both planted and naturally occurring
- tenure with respect to other natural resources, notably snails, mushrooms, and wildlife, especially game.

In Gyamfiase-Adenya the land is owned almost exclusively by extended families of the native Akuapem people on the basis of the *abusua* matrilineal or patrilineal kinship principle. Under this arrangement, members of the land-owning group have free access to the land. They may grant it out permanently or temporarily to others, including a growing number of tenants. A similar situation prevails in portions of Amanase-Whanabenya which are settled by offspring of Akuapem migrant-settler farmers who bought the land from Akyem/Akim people.

In other areas of Amanase-Whanabenya and in Sekesua-Osonson, the land is owned respectively by offspring of Shai/Siade and Krobo migrant-settler farmers. They bought the land also from Akyem people, according to the unique *huza* arrangement.

From about the middle of the nineteenth century, Krobo farmers sought new lands through an arrangement whereby individuals pooled their financial resources through companies or cooperative groups. In the Dangbe language of the Krobo and Shai, such a company is called *huza*. It was organized for the purpose of collective bargaining for land. After its purchase, the land was divided into privately owned longitudinal strips, *zugba* or *zugbakpo*, and shared proportionately among the company members according to each one's financial contribution towards the group land purchase. Commonly a homestead is constructed at the base of each of the parallel longitudinal or linear *zugba*. From the base, farming proceeds through the traditional system of land rotation, in the same general direction along the *zugba*. Subsequently, other migrant farmers, including those from Akuapem, apparently adopted the novel "company" concept.

A *zugba* is regarded as the original buyer's private estate, which he may dispose of howsoever. However, if the buyer dies intestate, i.e. without specifying the manner of treating the *zugba*, in accordance with the patrilineal custom of Krobo people, the eldest son of the deceased inherits it. It may also be shared among the sons of the deceased, with the proviso that it remains ancestral family property as in the matrilineal system of some Akuapem and other Akan people. As ancestral family property, an inherited *zugba* may not be alienated, except under dire circumstances. In effect, the inheritors are only trustees. They may, as in the case of Akuapem people, grant out inherited land to non-family members, especially in tenancy forms.

In Bongnayili-Dugu-Song, as in the whole of the Dagbon traditional area to which Bongnayili-Dugu-Song belongs, land is generally held in custody for the local people by chiefs and heads of clans and families under the overlordship of the Ya-Na, the supreme ruler of Dagbon. Among the chiefs under the Ya-Na are paramount ones including the Voggu, the principal custodian in Bongnayili-Dugu-Song. In each of the various communities, a *tindana*, *tindaanama*, or *tindamba* performs purely spiritual or religious functions over land. The local people have free access to land through their chiefs and heads of clans and families. Strangers, outsiders, or non-community members may have access by grants, which, traditionally, require only token recognition of the granting authority by a presentation of kola nuts and farm produce.

In Nyorigu-Binguri-Gore where the institution of chieftaincy is much less developed, the chiefs have virtually no control over land. The land is held and controlled on behalf of the community of owners by clans and families who do so under the spiritual oversight of a *tindana*. Through the custodians, access to land is granted to members of the community owning it, and to others upon a token offer as in the case of Bongnayili-Dugu-Song.

Thus, traditionally, access to land is achieved in two ways in all the sites. The first, the most fundamental, is based on kinship. It involves no payment, since the kin or members of the family owning the land enjoy free usufruct. The second is grants by groups owning land, which, typically, involve payment in cash or kind, either token or substantial. It, thus, is by contract (customarily unwritten) between landlord and tenant. However, access to land may be gained through the modern government, which has powers for compulsory acquisition.

Regarding tree tenure, traditionally economic trees are owned in common by those on whose land the trees occur. Following the imposition of British colonial rule, the government became the trustees and *de jure* owner of commercially valuable timber species, which are found almost exclusively in southern Ghana. Recently, in the post-colonial era, the law that vests ownership of such commercial trees in government is undergoing review.

Mushroom, snails, and other wildlife are generally treated as communal property. Usually owners of the land where such resources are found enjoy the right of first access.

Tenurial practices and biodiversity in southern Ghana sites

Tenancy versus owner-occupied management

In Gyamfiase-Adenya, tenancy, which involves mainly sharecropping and rental units by migrant-settler Ayigbe farmers originating from Togo and from Ghana's Volta region, has become significant. Land units farmed on a tenancy basis are estimated to comprise no less than 50 per cent of the total number of farmed units, even though the land continues to be owned, overwhelmingly, by the native Akuapem people on the basis of their *abusua* kinship tenurial arrangement. The tenants tend to cluster in communities separate from those of the Akuapem landowners.

Observations that are reported elsewhere indicate less favourable biophysical conditions in migrant-tenant farmer localities. Indications include the following:

- a greater dominance of grass and non-flora species
- fewer trees
- more erosion, soil exhaustion, acidic soils, and crusted surfaces
- more of cassava/manioc (*Manihot utilisima*), a crop that is tolerant of poor edaphic conditions, and less of crops that require better soil and moisture conditions, e.g. cocoa (*Theobroma cacao*), cocoyam (*Xanthosoma* sp.), and plantain (*Musa paradisiaca*)
- less crop diversity and a greater trend towards cassava monoculture
- greater landscape denudation.

In Gyamfiase village, which is predominantly owner-occupied, there remains a relict forest surrounded by a biodiverse indigenous agroforestry zone, which deteriorates towards Kokormu, a core stranger-tenant-farmer locality occupied by the Ayigbe people (Gyasi, 1998).

Similarly, biodiversity generally appears to be less in the tenant units in Sekesua-Osonson and Amanase-Whanabenya, where tenancies are estimated to comprise 30 per cent and owner-occupied units 70 per cent of the total number of farmed units. This is particularly so in Amanase-Whanabenya. The immediate surrounding of Amanase town is severely deforested and cropped to cassava/manioc, in many cases almost to the total exclusion of other crops, by recent Ayigbe migrant-settler tenant farmers. It contrasts sharply with the less deforested and more biodiverse distant areas of Whanabenya and Aboabo farmed to a wider assortment of crops on an owner-occupier basis by the landlords, who are Shai and Akuapem people. Their forebears migrated in about 100 years ago for cocoa cultivation (Hill, 1963). In a similar pattern, of all the land-use types surveyed by PLEC in the three demonstration sites, owner-managed home garden agroforestries showed the greatest biodiversity, especially in Sekesua-Osonson (Table 18.1).

From the preceding, on the whole biophysical conditions including the diversity of flora are less favourable in tenant-operated areas. The situation might be so because, probably, tenants are, for the most part, allotted exhausted plots or those of an inferior quality that cannot support a high diversity of plants life. But perhaps a more plausible explanation lies in the overfarming of tenancy units. Overfarming is fuelled by what, in group discussions, tenants described as usurious fees. Many of them could not afford a yearly rental fee of ¢24,000 (US$3 approximately) to ¢120,000 (US$15) per acre, or ¢57,600 (US$7.20) to ¢288,000 (US$36) per hectare. Under sharecropping, typically a landlord takes a third of the maize crop and a half of cassava and all other major crops, in addition to retaining rights over economically valuable trees such as the oil-palm.

As a tenant remarked in Gyamfiase-Adenya, "Due to the fact that our [tenants'] share of the cassava is not enough to pay for the labour cost and to meet other demands, we have to continuously till that same piece of land so as to ensure some income on a regular basis." Similarly, another in

Table 18.1 Concentration of home gardens per compound house in non-nucleated linear settlements relative to nucleated settlements in PLEC study sites

		Home garden concentration (number of home gardens per compound house)	
Study site	Settlement name	Non-nucleated	Nucleated
Amanase-Whanabenya	Whanabenya-Nyamebekyere	1.55	
	Abenabo	0.93	
	Obongo	1.15	
	Aboabo		0.75
	Amanase		0.85
	Aye-Kokooso		1.22
Sekesua-Osonson	Osonson Korlenya	2.20	
	Prekumasi	1.00	
	Siblinor	1.30	
	Sekesua		1.15
Gyamfiase-Adenya	Otwetiri	0.46	
	Kokormu		0.39
	Yensiso		0.70
Average		1.36	0.78

Source: A 1998/99 PLEC survey

Sekesua-Osonson commented, “Because they [the tenants] are compelled to hire the land for three years [i.e. short periods] they have to exploit it intensively to make up the hiring charges.” In the view of a Sekesua-Osonson landlord, “The tenants cut all the trees on the land to enable them to cultivate the whole land for higher yields since the higher their yield, the higher their shared part”, unlike the landowners, who “do not cut down all the trees”. Thus, it would seem that in order to make ends meet and ensure survival, at least in the short term, tenants are compelled by exacting tenancy obligations to over-farm the land. This undermines biodiversity.

About 40 years before the PLEC project, Ghana’s Ministry of Agriculture had noted the exploitative character of tenancy when it described renting of farmland as “the most destructive of all arrangements for holding land since the cultivator aims at obtaining a maximum of returns in the shortest possible time; and this objective transcends the need to adhere to good husbandry practices” (Division of Agricultural Economics, Ministry of Agriculture, 1962: 7). Earlier Varley and White (1958) had echoed this view. After observing environmental havoc caused apparently by migrant-tenant monocultural maize farming on rented land in the southern forest fringes (the forest-savanna transition zone) near Accra, the national capital, they concluded that “as a long-term policy this system [of transient migrant-tenant farming on rented land] is bad” (Varley and White, 1958: 92), because the tenants cut down most of the trees, carrying out intensive monocropping, "moving on when the land is exhausted" (Varley and White, 1958: 120). In this way, the tenants had rendered much of the area "completely useless for farming for many years to come" (Varley and White, 1958: 93).

In cropping the land, Ayigbe migrant-tenant farmers make extensive use of the hoe, which erodes biodiversity by damaging plant propagates or seed stock embedded in the soil.

In a PLEC sample questionnaire survey of farmers by household in the three sites, the question was posed, "What do landlords do that might inhibit conservation or optimal management of agrodiversity and associated natural/land/biophysical resources, such as trees?" With a percentage score of 89.7, high, exacting, or exploitative tenancy was the most frequently mentioned factor. It was followed by short duration of tenancies (64.1 per cent), uncertain nature of tenancy (46.2 per cent), and failure by landowners to grant permission for tree planting by tenants (35.9 per cent).

Another question posed was, "On balance, which one would you consider to be more favourable for conservation or optimal management of agrodiversity and associated natural/land/biophysical resources, such as trees – tenancy, or *abusua* (the kinship arrangement that entitles owners free use of land)?" The *abusua* system was cited as the most favourable by the majority of the respondents. Its percentage score was 91.7 compared to 8.3 for tenancy.

The kinship arrangement may be a better conserver of biophysical resources simply because it does not involve a compelling need to overexploit such resources to meet exacting tenancy obligations.

In Sekesua-Osonson, as in other areas settled by migrant Adangbe-speaking people, an aspect of resource tenure that favours biodiversity, most especially in home gardens is the *huza* system (Table 18.1). According to one account,

> The *huza* land is shared into individually owned longitudinal strips called *zugba*. Homes, or dwelling units, are constructed at the base of each *zugba*, giving rise to a linear settlement pattern that contrasts with the nucleated pattern typical of Akan areas. From the home area, farming proceeds in the same general direction along the *zugba*, uninhabited by other dwelling units, unlike what frequently happens in built-up nucleated settlements. Gardens commonly of the agroforestry type, are developed within a few hundred metres from the *zugba* home.
>
> The home garden agroforestry type contains virtually all the varieties of crops found in the non-home garden type and more. (Gyasi, 2002: 250–251; see also Chapter 12 of this book)

In Gyamfiase-Adenya, rights of harvesting fuelwood or firewood occurring naturally in a farm popularly belong to the farmer, whether the person is the landowner or not. As the arrangement benefits landowner and tenant alike, it is reasonable to expect it to favour tree conservation.

A contradiction

However, the preceding indications that tenancy exerts a less favourable effect on biodiversity and that it may in fact destroy biodiversity should be accepted with caution because they appear to be contradicted by the findings of a major PLEC survey of home gardens carried out in Gyamfiase-Adenya in 1998.

Among the food-crops farms invariably operated by migrant Ayigbe settler-tenants (including home gardeners) through land rotation in the bush away from the compound house, the PLEC team noticed a sizeable number in which the food crops grew alongside conserved trees, especially young ones, a practice encouraged by PLEC. This unexpected situation was probed further by a closer study of the floral composition of four of the tenant farms in which trees are conserved *in situ*, and by discussions with the operators of the farms, who may own the trees either exclusively or jointly with the landlord. Table 18.2 shows the diversity of plants in one of the studied farms, which is managed by C. K. Avume, a tenant farmer, at Otwetiri. The crops are dominated by cassava and maize, which are raised on a sharecropping basis. The 15 species of sapling/tree encountered are used variously as firewood, medicine, and constructional material.

Tenants cited growing scarcity of fuelwood associated with deforestation as the principal reason for the practice of tree conservation. Trees are harvested regularly for fuelwood. Other reasons are that trees serve the useful purposes of providing the following: medicine, supplementary food, wood for carving, fencing, and house construction, and mortar and pestle for pounding *fufu*, a popular local meal (Gyasi, 1999).

Contrary to expectation, the tenants reportedly started conserving trees spontaneously independently of the nearly four-year-old PLEC campaign for tree conservation in the Gyamfiase-Adenya demonstration site. This finding would seem to suggest that there is, inherent among tenants, environmental consciousness, which policy should recognize and use as a basis for biophysical conservation planning.

Situation in northern Ghana sites

In Bongnayili-Dugu-Song and Nyorigu-Binguri-Gonre sites, as in the rest of northern Ghana, the basic medium of access to farming land is through the kinship arrangement. Just as in southern Ghana, through this arrangement members of the land-owning group enjoy free usufruct. Others may be granted farming rights, but payment, if any, is only token, unlike the practice in southern Ghana, which involves high tenancy fees.

In Bongnayili-Dugu-Song and Nyorigu-Binguri-Gore, because tenancy is not very well developed, it can hardly be expected to have a profound impact on the biodiversity. But certain other tenurial aspects appear to do so. They include those

Table 18.2 Plants in a farm of C. K. Avume, a tenant

Crop				Other plants (sapling/tree)			
Local name in Twi	Common English name	Botanical name	Use	Local name in Twi	Common English name	Botanical name	Use
Bankye	Cassava/manioc	*Manihot esculenta*	F	*Pepea*	?	*Phylanthus discordens*	Fu
Aburow	Maize/corn	*Zea mays*	F	*Emire*	?	*Terminalia ivorensis*	T
Abe	Oil-palm	*Elaes guineensis*	F	*Ofosow*	?		Fu
Mako	Pepper	*Capsicum annum*	F	*Osese*	?	*Horralhaena floribunda*	F/C
Ase	Cowpea	*Vigna unguiculata*	F	*Osensrema*	?	*Neuboldia laevis*	M
Mango	Mango	*Mangifera indica*	F	*Owudifo akete*	?	*Anthocleista vogelli*	M
				Okronoo	Red-flowered silk cotton tree	*Bombax buonopozense*	M
				Awonwee	?	*Olax subscorpioidea*	M
				Opanpan	?		Fu
				Osena/yooye	Velvet tamarind	*Dialium guineese*	F/O
				Nyamedua	Pagoda tree	*Alstonia boonei*	M
				Akakapenpen	?	*Rauvolfia vomitora*	M/Fu
				Agyama	Christmas bush	*Alchonea cordifolia*	M
				Odwen	Camwood	*Baphia nitida*	M
				Bronyadua	Brimstone tree	*Morinda lucida*	M

F – Food; M – Medicinal; Fu – Fuelwood; O – Other use; T – Timber; C – Carving; ? – Not known.

relating to trees, which are highly valued because of their religious significance and utility as a source of the following items, among others:

- condiments and oil for cooking and other purposes
- energy for cooking
- income for especially females
- rafters for construction.

According to Agyepong *et al.*, customarily, "the right to harvest, collect, fell, and otherwise use trees generally rests with the custodian, who is the chief or the *tindamba*, except in the case of firewood and certain fruits from naturally propagated trees, which may be harvested by anybody from the community. In some cases, fruits may be shared between the custodians or the chief and the farmer who works on the land." (Agyepong *et al.*, 1999: 256–257; see also, Agyepong *et al.*, 1993)

In Bongnayili-Dugu-Song, there is a custom whereby "chiefs" are specially appointed purposely to oversee exploitation of exceptionally useful trees, for example the *dawadawa*, *Parkia* sp., a popular source of condiment and female income, whose chief is called Dawadawa Chief. Fruits of the *dawadawa* are harvested only with permission of the chief. By facilitating policing of trees, the custom of "Tree Chiefs" helps to conserve floral diversity. In Nyorigu-Binguri-Gore the relevant authority is vested in the *tindana*.

Tenurial norms also require the retention and nurturing of economic trees that sprout naturally in farmlands. In 2000, in a survey of 30 households in each of the two demonstration sites, four endemic tree species were consistently found to be retained in farming fields averaging 2.8 ha (seven acres) per household (Table 18.3).

Table 18.3 Endemic economic tree species retained in farms in Bongnayili-Dugu-Song and Nyorigu-Binguri-Gore

Tree species	Average number per household	
	Site 1: Bongnayili-Dugu-Song	Site 2: Nyorigu-Benguri-Gore
Shea, *Vitellaria paradoxa*	64	9
Dawadawa, *Parkia biglobosa*	10	4
Baobab	1	1
"Kapok", *Ceiba pentandra*	1	1
Average	19	3.8

Note: The far smaller number of trees per unit area in Nyorigu-Benguri-Gore is indicative of the severe population-induced deforestation there and in the Bawku area and the Upper East region as a whole.

Other tenurial aspects in southern and northern Ghana sites

In the sites within both southern and northern Ghana, a relevant tenurial aspect is the traditional perception of land by its owners as a sacred ancestral property. This perception has positive implications for biodiversity and biophysical status in general because, as a sacred ancestral property, land may not be alienated, but rather held and managed sustainably so as to ensure its availability for future generations.

Having a similar favourable positive implication for biodiversity is the special protection traditionally accorded to certain species of trees and groves of forest because of their being perceived as the abode or embodiment of gods and ancestral spirits. As such, biotic resources of this kind are placed under the special custody of an earth priest or spiritual head, notably *tindana*, through whom accessibility may be achieved.

A last tenurial aspect having a positive implication for biodiversity is the enduring customary perception of any tree planted as exclusively belonging to the person who planted it. By guaranteeing that benefits from trees planted accrue directly to the planter, people are likely to plant on a sustainable basis, thereby enhancing the quantum and diversity of trees.

Conclusion

There are aspects of resource tenure that favour biodiversity. They include:

- the kinship landholding arrangement, which appears to be less exploitative of the land than usurious tenancies
- the special protection accorded to land, some species of trees, and groves of forest because of their being perceived as ancestral property, sacred, or exceptionally useful
- the treatment of trees planted as belonging exclusively to the planter.

The challenge is to recognize, nurture, and popularize these and other positive tenurial aspects as part and parcel of the process of enhancing biodiversity conservation.

REFERENCES

Agyepong, G. T., J. S. Nabila, E. A. Gyasi, and S. K. Kufogbe, *Aspects of the Wider Spatial Context of CIPSEG*, UNESCO Co-operative Integrated Project on Savanna Ecosystems in Ghana: UNESCO-CIPSEG, A Baseline Study, Legon: Department of Geography and Resource Development, University of Ghana, 1993.

Agyepong, G. T., E. A. Gyasi, and J. S. Nabila with S. K. Kufogbe, "Population, land-use and the environment in a West African savanna ecosystem: An approach to sustainable land-use on community lands in northern Ghana", in B. S. Baudot, and W. R. Moomaw, eds, *People and their Planet: Searching for Balance*, London: Macmillan, New York: St. Martins's, 1999, pp. 251–271.

Division of Agricultural Economics, Ministry of Agriculture, "A report on a survey of land tenure systems in Ghana", *Ghanaian Bulletin of Agricultural Economics*, Vol. 2, No. 1, 1962, pp. 1–15.

Gyasi, E. A., "The adaptability of African communal land tenure to economic opportunity: The example of land acquisition for oil palm farming in Ghana", *Africa*, Vol. 64, No. 3, 1994, pp. 391–405.

Gyasi, E. A., "Land tenure system and traditional concepts of biodiversity conservation", in D. S. Amlalo, L. D. Atsiatorme and C. Fiati, eds, *Biodiversity Conservation: Traditional Knowledge and Modern Concepts*, Accra: Environmental Protection Agency, 1998, pp. 16–23.

Gyasi, E. A., "Claim that tenant-farmers do not conserve land resources: Counter evidence from a PLEC demonstration site in Ghana", *PLEC News and Views*, No. 12, 1999, pp. 10–14.

Gyasi, E. A., "Traditional forms of conserving biodiversity within agriculture: Their changing character in Ghana", in H. Brookfield, C. Padoch, H. Parsons, and M. Stocking, eds, *Cultivating Biodiversity: Understanding, Analysing and Using Agricultural Diversity*, London: ITDG Publishing, 2002, pp. 245–255.

Hill, P., *The Migrant Cocoa-Farmers of Southern Ghana: A Study in Rural Capitalism*, Cambridge: Cambridge University Press, 1963.

Varley, W. J. and H. P. White, *The Geography of Ghana*, London: Longman, 1958.

19

Resource access and distribution and the use of land in Tano-Odumasi, central Ghana

John A. Bakang, William Oduro, and Kwaku A. Nkyi

Introduction

This chapter discusses factors affecting resource access and distribution and how this is linked to livelihood strategies, associated land-use forms or stages, and their diversity and heterogeneity indices in Tano-Odumasi, central Ghana (Map B). It focuses on some management diversity issues in the context of practice theory. According to Ortner (1984, as cited in Endre Nyerges, 1997), "practice theory seeks to explain the relation(s) between (individual) human action, on the one hand, and some global entity which we may call 'the system' on the other. Questions concerning these relationships may go in either direction – the impact of the system on practice and the impact of practice on the system."

Tano-Odumasi is a small town in the Sekyere West district of Ashanti region. It is situated on the main Kumasi–Mampong road. It has a population of about 3,800 people in 1,350 households, with an average of seven members each, housed in 512 compounds. The traditional inhabitants are Ashanti people. The people are generally subsistence farmers who strive to produce surpluses for the market. They do not use chemical fertilizers in their farming practice except in vegetable production. A major management practice is slash-and-burn. Others include bush fallowing or land rotation, all of which may involve mixed and inter-cropping, and *proka* or *oprowka*, a no-burn mulching practice. The hoe and cutlass are the main tools.

Methodology

The methodology involves usage of information accumulated from surveys in the form of a database, and of information from iterative dialogue sessions and interaction between a multidisciplinary team of scientists and collaborating PLEC farmers, all under the UNU/PLEC project work in Ghana. Specific approaches and techniques were general meetings, focused group discussions, informal interviews, and keen observation in the homes and in the fields of collaborating farmers during the period February 2001–February 2002.

Discussion of findings

In an attempt to identify and trace linkages between land access and distribution and indigenous management diversity in the form of land use, the findings of this case study are discussed under the following sections:
- land access and distribution
- land-use forms or stages
- access and distribution, associated land-use forms, and diversity.

Land access and distribution

Access implies the right to use or benefit from a productive resource, while control refers to the effective exercise of such rights (Berry, 1993). Control is the ability and willingness to exclude other users (Convery, 1995).

On the basis of this clarification, holders of resource-use rights fall into three broad categories, namely communal, landowning groups (family), and the individual.

Communal rights are exercised over communal property, which is now limited to the sacred grove and graveyards. For historical reasons linked to the origin of the town and accepted by the entire community, access and distribution of land is not uniform across the landowning groups. For example, the royal Asona clan, originally from Bomwire, who first settled in and named the town after their god, Tano, has more land than any other landowning group.

As is well known, in Ghana, among the Akans in general and the Ashantis in particular, access to land within the family is inherited with descent being traced along matrilineal lineage, while distribution among households and individuals is effected through a complex of determinants of household power relations, age, gender, and social status.

Individuals derive their access to land through membership of the land-owning group, inheritance, gifts, outright purchase, pledging, or renting. Rent is either in cash, or in kind by sharecropping whereby payment takes the form of sharing of the farm produce or of the farm itself between the landowner and the operator in varying proportions, locally called *abunu*, *abusa*, etc., depending upon the proportionate distributions. In a third form of sharecropping, the operator retains maize while cassava and the land revert to the landowner in a maize/cassava intercrop.

Cash renting of land is preferred to sharecropping for the production of vegetable crops such as tomato, garden eggs, and pepper.

In Tano-Odumasi virtually every household, including those of migrants, has access to land for subsistence through routes other than renting or sharecropping. For example, in a survey of 123 households only three had access to land through tenancy only (Table 19.1). For a majority of individuals (77 per cent) in this survey, tenancy provides the means for accessing additional land for cultivation beyond subsistence levels. Renting or sharecropping is the preferred mode of exercise of rights by landowners.

Land-use forms/stages

The PLEC Biodiversity Advisory Group (BAG) recognizes the following land-use forms or stages:

- house (home) gardens
- annual cropping
- agroforests
- grass-, shrub-, and tree-dominated fallows
- orchards
- native forests (Zarin, Huijin, and Enu-Kwesi, 1999).

Within these general land-use categories the common field types identified in Tano-Odumasi are described below. A summary of the relationship between the

Table 19.1 Land access categories in Tano-Odumasi

Access/ownership category	Male	Female	Total
Owner-operator only	21	4	25
Tenancy only	1	2	3
Combined owner-operator and tenancy	36	59	95
Total	58	65	123

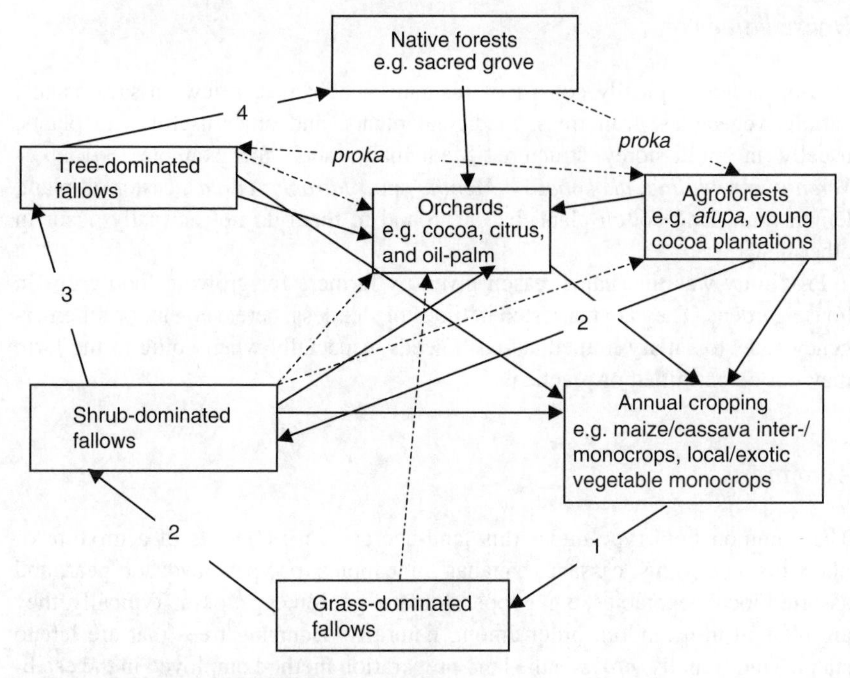

Figure 19.1 Management diversity and associated land-use forms in Tano-Odumasi

various land-use forms and specific field types and the management diversity associated with these is depicted in Figure 19.1.

Native forests

As the name implies, native forest involves non-cultivation. It essentially is a fallow at the peak of succession of the climax vegetation. It is characterized by the natural vegetation cover and is often the source of supply for medicinal plants, timber, and other non-timber forest products. Native forests are prime lands that are ideal for the development of cocoa plantations or the establishment of *afupa*, the main farm. Farmers, however, complain that cultivable native forests are hardly available in Tano-Odumasi. A relic is the sacred grove, an abandoned old settlement site of ancestral graves that is never to be cultivated. To protect the sacred grove from a growing encroachment, trees have been planted around it with the support of UNU/PLEC.

House gardens

House gardens typically consist of plantains, cocoyams, a few cassava stands, various vegetables, fruit trees, medicinal plants, and other useful wild plants, usually in multi-storey structure. Medicinal plants, for example, *onyono – Veronia amygdalina*, *akokobesa – Mentha* spp., *Jatropha curcas*, castor oil plant, lemon grass, and cotton plant that are found in them do not normally occur in other fields.

Proximity was the major reason given by farmers for growing food crops in house gardens. They are harvested in times of sickness, bereavement, or for emergency sales to satisfy immediate cash needs, especially when going to the farm may not be permitted or practical.

Agroforests

The common field type under this land-use category consists of a mixture of plantains, cocoyams, cassava, bananas, pineapples, pawpaw, avocado pear, and assorted local vegetables, e.g. pepper, tomato, garden eggs, okra. Typically, they are planted in a random order among naturally occurring trees that are left to stand. Traditionally, *proka* is the land preparation method employed in the establishment of this field type.

Proka *land management practice*

As noted in Chapters 6 and 11, *proka* is an Akan word which literally means to rot and affect. It is a management practice that involves clearing native forests or tree-dominated fallows without burning. The plant biomass is allowed to decompose *in situ*, thereby adding organic matter and nutrients to the soil. It is not a land-use form or stage. It is a practice. This traditional method of land preparation usually begins in August with slashing of shrubs and small trees, and the selective felling or killing by fire of big trees. Further clearing continues in the dry season (December/January) to create the necessary spaces for planting while awaiting the onset of the rains in February/March.

Planting of crops, notably plantains, cocoyams, cassava, fruits, and vegetables, is done after the rains have set in. Traditionally, cocoyams grew naturally and profusely alongside other crops and plants without being consciously planted by farmers. This mix of crops, together with the trees deliberately left on the field by the farmer, results in what has been described elsewhere as traditional small peasant farms under the general land-use stage of agroforestry. It is also this mix of crops that farmers in Tano-Odumasi refer to as *afupa*.

The concept of afupa *as a livelihood strategy*

Afupa, "the real farm", is the old traditional small peasant farm in which every attempt is made to satisfy the subsistence or food security needs of the household. The relative importance of this must be viewed in relation to the sources of income of the people. As indicated in Table 19.2, 62 per cent of them depend solely on farming for their income. It is interesting to note that irrespective of size, all other fields with plant species compositions that do not correspond to the mix of crops described above (including the most popular maize/cassava intercrops or monocrop) are regarded as inferior or less important and not considered to be real farms.

When one is informed that a farmer has left for "his farm" without any further information, it means, to the people of Tano-Odumasi, that the farmer has actually gone to his real farm or *afupa*. However, when one is informed that a farmer has gone to "the farm", *afuom*, further questions need to be posed to ascertain which particular field is being referred to. This minor distinction in the words of the people has important household food security implications. It is embedded in the definition of *afupa* as "a field on which one can depend on to provide the food needs of the household, or a field that provides everything we need" (personal communication with Yaw Bio, a PLEC farmer of Tano-Odumasi).

Consequently, the traditional *afupa* has always been associated with the most fertile lands of the household, which, however, are now getting increasingly located further away from the settlement.

Another feature of *afupa* is its symbolic position. Generally, every family of the landowning group and, to a small extent, a few immigrants of good standing possess it.

Orchards

Under orchards, the different field types include cocoa, citrus, and oil-palm plantations in various stages of establishment. Cocoa plantations are usually developed on tree-dominated fallows or mature uncultivated forests because of the initial shade requirements of seedlings. Citrus and oil-palm plantations, on

Table 19.2 Sources of income in Tano-Odumasi

Source of income	Males	Females	Total
Full-time farming	38	38	76 (62%)
Non-farming	20	27	47 (38%)
Total	58	65	123

the other hand, are usually developed following total land clearing. They could be developed from grass-, shrub-, and/or tree-dominated fallows.

The orchards are usually interplanted with cocoyams and plantains depending on the stage of their development. In the case of citrus and oil-palm, other annual food crops, notably maize and cassava, are also interplanted until the closure of the canopies of cocoa and oil-palm. Thereafter, interplanting is possible only in citrus plantations.

Annual cropping

Distinct field types under this land-use category include mono- or intercrops of maize and cassava, and monocrops of vegetables (tomatoes, okra, garden eggs, peppers, and cabbages). The most common in this land-use category, however, is the maize/cassava intercrop. Monocrops of vegetables were found to be popular among younger farmers who practised market gardening in order to take advantage of the nearby urban market in the city of Kumasi. The cultivation of these vegetables, usually in valley bottom lands, is always associated with the indiscriminate use of pesticides, with or without the use of chemical fertilizers and/or supplementary irrigation.

Grass-, shrub-, and tree-dominated fallows

Fallows dominated by grass, shrubs, and trees are common in Tano-Odumasi. Grass-dominated fallows are usually fallow fields emerging from a period of continuous annual cropping.

With a sustained period of fallow (four to five years), shrubs develop and gradually shade out grasses to result in the shrub-dominated fallow.

Tree-dominated fallows usually result from abandoned small peasant agroforests, *afupa*, which begin the fallow cycle with shrubs.

The difference between the two shrub-dominated fallows lies in the higher preponderance of trees and other plant species in the fallow with the recent history of agroforestry.

The concept of good and bad trees

Although a new Forestry Law now recognizes the rights of the landowner to trees occurring on his/her land, certain farmer perceptions with regard to trees in the various land-use stages can affect farmers' attitudes towards nurturing trees on their land. Farmers in Tano-Odumasi believe that their crop plants perform differently in proximity with different trees, hence the concept of good and bad

Table 19.3 Good and bad trees

Local name	Botanical name	Farmer perceptions	
		Good/bad	Availability
Onyina	*Ceiba pentandra*	Good	Available
Kokoonisuo	*Spathodea campanulata*	Good	Available
Framoo	*Terminalia superba*	Good	Less available
Wama	*Ricinodendronh endelotu*	Good	Available
Otie	*Pycnanthusang olensis*	Good	Less available
Wawa	*Triplochiton scleroxylon*	Bad	Less available
Esa	*Celtis zankeri*	Bad	Less available
Okro	*Albuziagia*	Bad	Available
Denya	*Cylicodiscus gabunensis*	Bad	Less available

trees. "Good trees" generally enhance the yield of their crops while "bad trees" adversely affect yield. "Neutral trees" are also recognized.

This concept of good, bad, and neutral trees has implications for biodiversity since on-farm *in-situ* conservation and/or deliberately planting of trees in the field will be influenced by these perceptions. This becomes even more important when it is realized that these perceptions may not have arisen out of personal experience but from hearsay or tradition.

Table 19.3 gives examples of farmers' classification of trees according to this concept, obtained from discussions with some farmers. In general, farmers believe that tree populations have dwindled considerably, with some becoming endangered or less available, or even extinct.

It may be worthwhile finding out the extent to which particular tree populations are affected by farmers' perception of good and bad trees. Similarly, it may be interesting for agronomists to learn from farmers and to verify whether there are scientific bases for these perceptions.

Access to and distribution of land, associated land-use stages/forms, and biodiversity

Access and distribution

As discussed earlier, access to land and its distribution are unequal in Tano-Odumasi. This accepted unequal access and distribution exists among the land-owning groups, between and within families, and between indigenes and migrants.

Table 19.4 Access to land and associated land-use forms

Access routes	Land-use forms
Membership of landowning family	Agroforests (*afupa*), orchards, fallows (grass-, shrub-, and tree-dominated), annual (sharecropping, house gardens)
Outright purchase	Orchards (cocoa, citrus, oil-palm)
Cash renting	Maize, cassava, local, and exotic vegetables
Sharecropping	Orchards, annual cropping

Generally, members of the landowning group have more land than immigrants. Similarly, family heads and elders of the landowning groups generally have more land at their disposal than ordinary/younger members do. Farmers who have more land at their disposal tend to explore all the land-use forms identified in Tano-Odumasi (Table 19.4).

Outright purchases on a large scale are generally more common in orchards than in the traditional subsistence agroforestry, *afupa*. Where farmers were involved in annual cropping, chemical fertilizers are more likely to be employed in maintaining or improving soil fertility. They are also less likely to fallow their lands.

Access through cash renting is usually for annual cropping, especially vegetable production. Generally farmers who access land using this route are more often than not involved in commercial production for the market rather than for subsistence. Migrants, young and ambitious indigenes, and mostly salaried absentee farmers form the majority of this category.

Sharecropping is practised under two major land-use forms, namely annual cropping and orchards (cocoa, oil-palm). As noted above (Table 19.4), sharecropping of annual crops and also orchards is the preferred route of exercise of access rights by elderly members of the landowning groups, while outright purchasers sharecrop in orchards.

Land-use forms and biodiversity

Biodiversity values for various land-use forms in Tano-Odumasi are presented in Tables 19.5–19.11. These include diversity and heterogeneity indices for three plot sizes (Table 19.5), number of species in common among individual land-use forms for three plots sizes (Tables 19.6–19.8), and species similarity values within individual land-use forms for three plot sizes (Tables 19.9–19.11). The tables show comparably similar values among some land-use forms (agroforest, native forest, and house gardens). The similarity in values between the *proka* field and agroforest seems to support the report from farmers that the two differ only in the initial land preparation.

Table 19.5 Diversity and heterogeneity indices for various land-use forms (for three plot sizes) in Tano-Odumasi demonstration site

Plot area (m^2)	Land-use form	Diversity and heterogeneity indices					
		Margalef	Menhinink	Gleason	Shannon	Simpson	Brillouin
400	1	0	0	0	0	0	0
400	2	3.51	2.77	3.90	2.20	0.05	1.54
400	4	2.92	2.41	3.34	1.97	0.73	1.37
400	6	2.16	2.41	2.89	1.39	0	0
400	7	0	0	0	0	0	0
400	8	8.33	4.12	8.55	3.26	0.05	2.75
400	9	2.02	1.73	2.41	1.66	0.14	1.22
400	12	0	0	0	0	0	0
25	1	4.77	0.85	4.90	1.98	0.23	1.95
25	2	6.08	1.28	6.22	2.17	0.22	2.11
25	4	5.99	1.37	6.14	2.04	0.27	1.97
25	6	4.31	1.22	4.47	2.05	0.19	1.97
25	7	4.59	2.38	4.82	2.46	0.12	2.13
25	8	7.90	2.35	8.06	3.22	0.06	3.04
25	9	6.82	1.79	6.98	2.06	0.23	3.61
25	12	5.55	1.54	5.71	1.75	0.32	1.66
1	1	3.93	0.51	4.05	1.33	0.61	1.12
1	2	6.41	1.10	6.54	2.39	0.14	2.35
1	4	7.62	2.73	7.80	3.04	0.01	2.80
1	6	6.52	1.54	6.67	2.61	0.14	2.52
1	7	2.87	2.09	3.19	2.11	0.10	2.04
1	8	7.11	2.24	7.27	2.55	0.14	2.39
1	9	5.41	1.42	5.70	2.42	0.16	2.33
1	12	6.03	1.40	6.17	2.44	0.16	2.36

Note: Land-use codes for Tables 9.5–9.11:
1. Annual cropping; 2. Agroforestry; 4. Shrub-dominated fallow; 6. Orchard – plantation; 7. Orchard – trees; 8. Native forest (sacred grove); 9. House garden; 12. The special case of *proka* (agroforestry)*.
* It has been explained why *proka* is not a land-use form or stage but a method of land preparation employed in traditional peasant agroforests.

Table 19.6 Number of species in common among individual land-use forms in 20 × 20 m plots in Tano-Odumasi demonstration site

	Land-use forms							
Land-use form	1	2	4	6	7	8	9	12
1	0							
2	0	10						
4	0	1	8					
6	0	0	0	5				
7	0	0	0	0	1			
8	0	5	3	3	0	39		
9	0	0	1	0	0	0	6	
12	0	0	0	0	0	0	0	0

Table 19.7 Number of species in common among individual land-use forms in 5 × 5 m plots in Tano-Odumasi demonstration site

	Land-use forms							
Land-use form	1	2	4	6	7	8	9	12
1	40							
2	20	44						
4	9	12	20					
6	11	9	5	15				
7	12	8	8	5	21			
8	12	15	5	5	6	49		
9	23	23	10	6	12	13	45	
12	20	20	8	8	5	9	18	39

Table 19.8 Number of species in common among individual land-use forms in 1 × 1 m sub-subplots in Tano-Odumasi demonstration site

	Land-use forms							
Land-use form	1	2	4	6	7	8	9	12
1	36							
2	26	52						
4	16	25	43					
6	22	25	17	46				
7	4	9	7	6	10			
8	11	15	15	14	3	43		
9	16	23	13	16	4	9	36	
12	22	26	18	18	6	8	20	43

Table 19.9 Species similarity values within individual land-use forms in 20 × 20 m plots in Tano-Odumasi demonstration site

Land-use form	Land-use forms 1	2	4	6	7	8	9	12
1	–							
2	0	–						
4	0	11.1	–					
6	0	0	0	–				
7	0	0	0	0	–			
8	0	20.4	12.8	13.6	0	–		
9	0	0	18.2	0	0	0	–	
12	0	0	0	0	1.2	0	0	0

Mean ± standard error = 2.7 ± 1.2%

Table 19.10 Species similarity values within individual land-use forms in 5 × 5 m plots in Tano-Odumasi demonstration site

Land-use form	Land-use forms 1	2	4	6	7	8	9	12
1	–							
2	47.6	–						
4	30.0	37.5	–					
6	40.0	30.5	28.6	–				
7	39.3	24.6	39.0	27.8	–			
8	27.0	32.3	14.5	15.6	17.1	–		
9	54.1	51.7	30.8	20.0	36.4	27.7	–	
12	50.6	48.2	27.1	29.6	16.7	20.5	42.9	–

Mean ± standard error = 32.4 ± 2.2%

Table 19.11 Species similarity values within individual land-use forms in 1 × 1 m sub-subplots in Tano-Odumasi demonstration site

Land-use form	Land-use forms 1	2	4	6	7	8	9	12
1	–							
2	59.1	–						
4	40.5	52.6	–					
6	53.7	51.0	38.2	–				
7	17.4	29.0	26.4	21.4	–			
8	27.8	31.6	41.9	31.5	11.3	–		
9	44.4	52.3	32.9	39.0	17.4	23.4	–	
12	55.7	52.6	41.9	40.4	22.6	18.6	50.6	–

Mean ± standard error = 32.6 ± 2.6%

Summary and conclusions

Proka, *afupa*, and "good" and "bad" trees, the three indigenous farmer concepts identified in Tano-Odumasi, have important implications for scientists in understanding some of the forces of agro-ecological change in the landscape of smallholder farmers.

The concept of "good" and "bad" trees has implications for biodiversity since on-farm *in-situ* conservation and/or deliberate planting of trees in the field will be influenced by these perceptions. This becomes even more important when it is realized that these perceptions may not have arisen out of personal experience but from hearsay or tradition. A first challenge is to find out the extent to which particular tree populations are affected by farmers' perceptions of "good" and "bad" trees. Similarly, it may be interesting for agronomists in particular to learn from farmers and to verify whether there are scientific bases for these perceptions. The relevance of this new learning process (for both scientists and agricultural extension workers) stems from the realization that despite increasing evidence of the range of farmers' strategies for dealing with degradation (Shepherd, 1992; Tiffen, Mortimore, and Gichuki, 1994; Harris, 1996), indigenous knowledge and practices have often not been recognized as worthy sources of sustainable resource management practices (Bakang, 2000).

A second challenge is how policy makers should promote the concepts of *proka* and *afupa* as sustainable management practices in order to address the core problems of deterioration of soil quality, declining agricultural productivity, deforestation, and depletion of species associated with other management practices of smallholder farmers. The challenge is rendered daunting by the increasing population pressure on land which is slowly ebbing away indigenous sustainable practices of farmers, who accept that these practices are dying out.

REFERENCES

Bakang, J. A., "Resource relations and degradation: A case study of the Dagaaba of Upper West region, Ghana", PhD thesis, University of Reading. 2000, unpublished.

Berry, S., *No Condition is Permanent. The Social Dynamics of Agrarian Change in Sub-Saharan Africa*, Madison: University of Wisconsin Press, 1993.

Convery, F. J., "Property rights and tenure systems", in *Applying Environmental Economics in Africa*, Washington, DC: World Bank, 1995, pp. 77–82.

Endre Nyerges, A., "Introduction – The ecology of practice", in S. H. Katz and A. Endre Nyerges, eds, *Food and Nutrition in History and Anthropology. Vol. 4: The Ecology of Practice. Studies in Food Crop Production in Sub-Saharan Africa*, Amsterdam: OPA, 1997, pp. 1–38.

Harris, F., "Intensification of agriculture in the semi-arid areas: Lessons from the Kano closed-settled zone, Nigeria", IIED Sustainable Agriculture Programme Gatekeeper Series, 1966, No. SA59.

Shepherd, G., *Managing Africa's Tropical Dry Forests: A Review of Indigenous Methods*, London: ODI, 1992.

Tiffen, M., M. Mortimore, and F. Gichuki, *More People, Less Erosion: Environmental Recovery in Kenya*, Chichester: John Wiley & Sons, 1994.

Zarin, D. J., G. Huijun, and L. Enu-Kwesi, "Methods for the assessment of plant diversity in complex agricultural landscapes: Guidelines for data collection and analysis from the PLEC Biodiversity Advisory Group (PLEC-BAG)", *PLEC News and Views*, No. 13, 1999, pp. 3–16.

20

Women environmental pacesetters of Jachie

Olivia Agbenyega and William Oduro

Introduction: Jachie women – The pacesetters

In most societies, women, when given the opportunity, have always played important roles as pacesetters. With the appropriate encouragement they are able to participate in activities that promote their welfare and the welfare of their communities. Traditionally, where it would ultimately benefit their families and communities, women have been known to engage in environmental activities that require a lot of sacrifice. The activities of the Jachie women's group are remarkable examples of what women can do when faced with challenges.

Jachie is one of the towns located in the dry semi-deciduous forest ecosystem of Ghana (Map B). In recent times, the semi-deciduous forest ecosystem has undergone a lot of changes. The changes include severe land degradation due to growing population and its attendant increases in the use of natural resources. This situation poses a threat to future food production (Shiferaw and Holden, 2000). Even so, in villages such as Jachie, a few patches of forest lands still in their pristine state remind us of what is left of the past. The women of Jachie are determined to protect and develop these remnants of forest lands for themselves, their community, and for future generations.

The PLEC women's project at Jachie was launched in 1993 to improve agrobiodiversity. The process of agrobiodiversity conservation was to be accompanied by the development of village-level institutions to serve as a basis for sustainable management of natural resources. It was realized during the creation of the project that the success of any intervention in regenerating the degraded farmlands would depend on the active involvement of women groups. The degradation of land through deforestation is a common problem affecting all, but has a more devastating effect on women.

The use of women as agrobiodiversity conservators is based on the belief that women are the main users of natural resources. Indeed, women have been shown to be the principal gatherers and users of forests and common land resources (Falconner, 1992). Products gathered are either used for subsistence or are processed into marketable commodities.

Trees and forests in Jachie are a source of a variety of products that are well integrated in the livelihoods of the women. In addition to farming, women keep small populations of livestock such as local fowls, ducks, goats, and sheep. They occasionally collect medicinal plants for trade. These biological items formed the basis for the planning and implementation of agrobiodiversity conservation in the village.

Jachie has a social situation which is typical of villages close to cities. Most men in this village have migrated to the nearby city of Kumasi to earn a living. This has created a new spatial division of labour and the formation of split households. Women farmers in this village now serve as caretakers of the home and natural environment. The women of Jachie, as is the case in other areas, make important contributions to agricultural production (Kalinda, Filson, and Shute, 2000). They have expanded upon traditional gender-based systems of managing their farms and landscapes. They have incorporated new forms of labour and responsibilities into their own activities. These roles were formerly part of the men's domain. This has made them leaders in the maintenance of agrodiversity. Faced with the task of ensuring the sustainability of their environmental resources, the women readily adopted the ideas presented by PLEC.

Genesis of the women farmers' group at Jachie

The PLEC project placed a lot of emphasis on women playing key roles in planning and implementing land-based activities. During the initial stages of PLEC, research scientists organized several visits to the village. They held consultations with the existing traditional and political groups, chiefs, elders, assemblymen, and village development committee members. This facilitated the establishment of an excellent working relationship between the village community and the scientists. A lot of progress was achieved through community forums, awareness creation, and community mobilization. In order to understand local village dynamics and specific issues relating to women, several hours were spent with the women in their homes, farms, and at social events such as funerals and weddings.

A situational analysis facilitated the identification of people who were ready to share knowledge. This was important because before the inception of PLEC activities, no previous attempt had been made in Jachie to get the women together. Therefore, to form the group, women who were articulate, with good communication skills and were willing to accept the responsibility of organizing group activities were selected to provide the initial support. These women were identified as potential members and leaders, and were assigned the role of educating and soliciting the participation of the other women in the village. To prevent the development of a dependency situation, the women were given the opportunity to show how they could contribute to the success of the group.

One individual who was instrumental in the formation of the Jachie women's farmer group is Madam Cecelia Osei (Plate 6). She was part of the leadership which provided the link between the researchers and the women. It has been through her tireless efforts and commitment that the group has managed to make an impact in Jachie.

Through the efforts of Madam Osei and her executive members, the association of PLEC women farmers was inaugurated in the community in 1992. The leaders were given training in organization and communication skills. The women leaders on their part conducted village-level meetings to increase women's awareness of project activities. An important outcome of the establishment of the PLEC Women Farmers' Association was the recognition of the women as important stakeholders in the management of the natural resources. In this programme, diversity was the critical issue in using scarce resources efficiently.

One of the initial measures that was instituted to help avoid further deforestation and to restore degraded areas was to control and sustain the current level of fuelwood production. Trees were planted to improve crop yields through soil fertility enhancement. As the project management gained working experience, innovative measures were introduced. These were specifically designed to increase agricultural diversity and participation by more women. The approach has been to work together with the women farmers' group to establish model outreach experimental farms that can meet most of the food requirements of a family with scarce capital and land resources.

Achievements

Overview

Through their own efforts and with support from PLEC researchers, the women of Jachie have accomplished a lot in terms of social and economic benefits. They report that the group has helped them in various ways to acquire skills in nursery management and livestock production. Through their activities they now understand the value of biodiversity. The women have also acquired some knowledge on the harmful effects of burning off vegetation after clearing their lands, and this has led to a significant reduction in the incidence of bushfires. In the last two years they have not experienced any bushfires. The general consensus is that the women who are members of the group have acquired the ability to manage their financial affairs in more beneficial ways.

Active participation of the women in the project

The level of participation in project activities has seen a steady rise over the years. The group, which started with 25 members, now has 432 members, distributed among Jachie and five surrounding communities, namely Akwaduo, Swedru, Apiankra, Nnuaso, and Homebenase.

Woodlot establishment

A major problem faced by the women was the shortage of fuelwood. This was a consequence of forest depletion. To overcome this problem, the women established a 10 ha plantation of teak (*Tectona grandis*) and *Cidrella* sp. in 1993. The teak was planted to provide electricity poles for the village and also money for the group from the sale of electrical and building poles. The *Cidrella* sp. was harvested early to meet fuelwood needs. Today the woodlot has become one of the major achievements of the group. Increased availability of adequate electric poles and fuelwood has established support for the project. Some of the teak trees have been planted on the land of the local Anglican school. This has reduced the problem of encroachment of the land belonging to the school. The trees also provide shade for children and adults in the village.

A further benefit of the establishment of the woodlot has been the savings in time previously spent in fuelwood collection. Prior to this the women had to spend about four to five hours collecting one head-load of wood (approximately 30 kg). Now the same quantity of fuelwood is obtained in two hours. Cooke (1998) reported that as environmental goods such as fuelwood and fodder become scarce, rural households in developing countries spend more time in their collection.

The popularity of the project with the women is a result of the excellent interaction between the researchers and the women. Together they have used participatory rural appraisal (PRA) techniques in the preparation of detailed micro plans. The women themselves determined which species was to be planted. They gave very high ranking to species used for fuelwood, such as *Margaritaria discoidea* (*pepia* – local Akan-Twi name) and *Celtis milbracdii* (*esa*), *Dacroydes klineana* (*akyia*), *Ficus exasperata* (*nyankyene*), *Funtumia elastica* (*fruntum*), *Albizia zigia* (*okro*), *Rauvolfia vomitoria* (*kakapenpen*), and *Hevea brasiliensis*, pararubber.

Nursery establishment

In 1993 the group established a nursery to produce tree seedlings. Groups of between four and 10 women were organized to produce the seedlings. The women provided the raw materials, including seed and site, while the polythene bags were provided by the project. In its first year, the PLEC women association raised nearly 20,000 seedlings. These were eventually used in their agroforestry and watershed programme. The number of seedlings increased over 80 per cent between 1996 and 2001. The nursery project and sales of seedlings have led to a sustained increase in village-level incomes and reduced the overall economic dependence on farm produce.

Agroforestry establishment

Traditionally women in the project area obtained their fuelwood from farmlands. The women's need for fuelwood was used to ensure protection of the plantations and to secure women's participation in agroforestry activities. Agroforestry on farmlands has been proposed as a viable approach to the problems of fuelwood shortage and decreasing soil fertility.

Institution of formal savings

Experience elsewhere (Tuteja, 2000) has shown that an increase in a women's income does not necessarily provide them with better development opportunities. This is due to the constraints of existing social structures and cultural practices found within the village and family. Traditionally, women have very little control over their incomes. To combat these and other problems and also to provide the women with training in understanding the operations of national banks and post offices, the Jachie women were encouraged to open savings accounts in their local "rural" banks. The PLEC women association has used its savings to:

- purchase planting materials
- hire labour
- purchase animal stock for rearing, and nursery materials for raising seedlings to enhance agrobiodiversity.

This has given women self-confidence as they have control over their own money.

Participation in workshops

Various workshops have been organized for the women to help them acquire skills to help in the management of their resources. They have been trained in techniques such as budding and grafting so that they could start raising horticultural and ornamental plants. The women have also been trained in proper methods for pruning. They have been exposed to scientific techniques in rearing snails, mushrooms, and beekeeping to enhance agrobiodiversity. All these help increase their income and also help reduce pressure on the wild species and ecosystem. Interest was generated during these training programmes through singing and drama. Women leaders have also played active roles in regional and international workshops for scientists and influential policy-makers. This has raised the status of the PLEC women farmers. They are highly admired and always acknowledged during formal functions in the community.

Involvement in watershed design and management

With the support of scientists, the women have designed watershed management strategies in the community. Encouraging results of watershed management programmes in India (Samra and Sikka, 1998) suggest that the adoption of an incentive-based and community-driven bottom-up approach for managing degraded watersheds on a sustainable basis is a promising approach. A community pond, *denkyemni*, is being restored for aquatic biodiversity and fish production. The women have inspired the community to come together to construct drainage structures for the pond with community resources.

Conservation of rare foods and medicinal plants

A rare-food-crops and medicinal-plants arboretum has been established. Together with others in the community, the women have collected endangered species for protection, propagation, and transplanting. In line with this, the researchers assisted them to label, document uses, and establish the arboretum. A sacred grove that was being threatened was restocked and is being managed to retain its natural vegetation.

Involvement of children in conservation activities

The women have been educating school pupils to change their perception that farming is a non-lucrative occupation. Promotional materials such as T-shirts with the inscription "conserving the land for future generations" have been produced. The children have been taught that planting crops is not the only land-based activity – there is also planting of trees, which has the potential to provide useful products and services. In addition to this, some of the skills the women have acquired from the scientists on processing of food crops are being passed on to the children.

Challenges

The success achieved by the women in PLEC has brought its own problems. Women who had hitherto lived sheltered lives, largely untouched by mainstream development, are excited about their new opportunities and want to make the most of these opportunities. However, some men in the village feel threatened by the developing financial freedom of the women. Other problems faced include transportation during educational campaigns. Another big constraint is the presence of termites, which destroy their food crops and their beehives. Despite all this, the women are confident that while they continue to collaborate with the PLEC researchers most of these constraints will be managed.

Conclusion

Pacesetters will always be pacesetters, and the women of Jachie believe in lighting the path for others to follow. This they have done by promoting biodiversity management. In the short span of seven years, PLEC has made tremendous progress in the confidence-building process as women now participate in village meetings. Nationally and internationally the project has attracted the attention of both visitors and the media. Their activities have been highlighted on national television as a successful example of development projects managed by women. An achievement worth mentioning is that the District Assembly Unit Committee elections of December 1998 brought about political changes in the village leadership. For the first time a woman, the leader of the PLEC Women Farmers' Association, was democratically elected as a member of the Unit Committee. As the major group of farmers, collectors of fodder, firewood, and other produce, the women continue to play a major role in sustaining agrobiodiversity on individual farm and village lands.

REFERENCES

Cooke, P. A., "The effect of environmental good scarcity on own farm labor allocation: The case of agricultural households in rural Nepal", *Environment and Development Economics*, Vol. 3, No. 4, 1998, pp. 443–469.

Falconner, J., *The Importance of NTFPs in the Rural Economies of Southern Ghana: The Main Report*, Report to Forestry Department, Accra and ODA, London, 1992, unpublished.

Kalinda, T., G. Filson, and J. Shute, "Resources, household decision making and organization of labour in food production among small-scale farmers in southern Zambia", *Development Southern Africa*, Vol. 17, No. 2, 2000, pp. 165–174.

Samra, J. S. and A. K. Sikka, "Participatory watershed management in India", in *Towards Sustainable Land Use: Furthering Cooperation Between People and Institutions*, Vol. II, *Proceedings of the International Soil Conservation Organization*, Bonn, Germany, 26–30 August 1996. *Advances in Geoecology*, No. 31, 1998, pp. 1145–1150.

Shiferaw, B. and S. T. Holden, "Policy instruments for sustainable land management: The case of highland small holders in Ethiopia", *Agricultural Economics*, Vol. 20, No. 3, 2000, pp. 217–232.

Tuteja, U., "Contribution of female agricultural workers in family income and their status in Haryana", *Indian Journal of Agricultural Economics*, Vol. 55, No. 2, 2000, pp. 136–148.

Part IV

Conclusion

21

Lessons learnt and future research directions

Edwin A. Gyasi

Introduction

From the preceding chapters, it is clear that the varied traditional ways of managing agrodiversity in West Africa have lessons for sustainable conservation of biodiversity and associated natural resources of the land or environment.

This chapter provides a synthesis of the key lessons relating to cultivation or management and organization of agrodiversity, and those relating to research methodology. Both aspects bear on official policy.

Management and organization of agrodiversity

The areas used for agriculture by the smallholder traditional farmers are characterized by high biodiversity of plants – both spontaneous and introduced. This situation is both a result and a cause of a complex, diverse traditional management and organizational arrangements, which could usefully inform development of modern agricultural improvement packages that enhance yield and conserve biodiversity simultaneously.

Commonly, highly biodiverse patches of forest are conserved because, like water bodies and land itself, they are traditionally perceived as sacred or as ancestral property. As such, they must be preserved or used sustainably and passed on to posterity. The perception of land as sacred ancestral property underlies the customary land tenure arrangement based on kinship, which appears to be less exploitative of biophysical resources than the increasingly widespread tenancies. This philosophy, like the traditional practice of placing highly valued tree species under the custody of an earth priest, *tindana*, is a dying ecologically positive one that warrants reactivation to encourage conservation of biophysical resources.

By not burning cleared vegetation, but rather using it for mulching, the traditional practice called *oprowka* or *proka* conserves plants, seeds, and soil microbes in addition to enhancing soil nutrients by humus. It may be improved by scientific studies with a special emphasis upon control of weeds and pests, which farmers consider to be major disadvantages of the practice.

Bush fallowing, the most popular traditional system of food cropping, is often condemned for its extensive and less than expected productive character. However, by periodic fallowing it regenerates and preserves a diversity of plants. Policy, therefore, may encourage it by supporting research focused upon higher-yielding biodiverse rotational cycles under different ecological conditions.

The still widespread traditional practice of intermixing crops, often among trees deliberately left *in situ* in farms, sufficiently conserves biodiversity and enhances ecological services as to warrant its encouragement. In some of the areas, notably those settled by migrant Krobo people, the mixed cropping has been developed into a home-centred agroforestry system embodying a rich diversity of ecologically and economically useful plants. The system has positive implications for food security and rural livelihoods. Its biodiverse character and proven economic viability points to its potential as a more viable alternative to the generally failed attempts at promoting exotic agroforestry systems. Further research, especially into spacing of crops relative to trees and into the scientific basis of the farmer perception of the agro-ecological functions of the trees, would enhance realization of the promise of the traditional agroforestry systems.

Another traditional agrodiverse system having positive implications for biodiversity and ecological services as well as food security, and therefore deserving of policy attention, is the management of yams by:

- planting them within agroforestry systems
- staggering their harvesting by lifting only one or two matured tubers from a bunch at a time, leaving the rest *in situ* in the ground for harvesting later, which has positive implications not only for food supply but also for seed stock
- leaving unharvested in the ground small nodules so that they may regenerate *in situ* in a process that may continue for years;
- live staking the climbers by using trees specially left standing for that purpose
- collecting yams from the wild and domesticating them in shrines before their transfer for cultivation in farms as part and parcel of the plant domestication process.

The pivotal role played by female farmers in the conservation of traditional varieties of rice at Gore, and the able stewardship of PLEC activities by females at Jachie, suggest that given control and other requisite resources, women would register a much greater impact within the agricultural sector, which is widely seen as holding a fundamental key to poverty reduction and the development process in developing countries (Todaro, 1989, 2000).

Often, conservation and development (i.e. that process of livelihood improvement) are seen as being at odds with each other, or as mutually antagonistic. However, the economic ventures (notably honey beekeeping) successfully carried out in conserved forests and fallow areas, and the successful woodlot ventures, demonstrate that there are ways of generating from conservation economic benefits (on top of the ecological ones) for local people, above all the farmers.

Research methodology

By successfully bringing together scientists from a diversity of academic backgrounds into a functional team, whose collective efforts have resulted in this book, the research methodology demonstrates the efficacy of interdisciplinary or multidisciplinary approaches.

By sustainably teaming up scientists from three universities (the University of Ghana, Kwame Nkrumah University of Science and Technology, and the University for Development Studies), the research methodology demonstrates the feasibility of generating positive research synergies through institutional collaboration by networking.

The richness of the information base of the book underscores the effectiveness of participatory procedures involving:

- collaborative work by scientists, farmers, policy-makers, extension agents, and other environmental stakeholders
- tapping and building upon traditional farmer resource management knowledge, especially through expert farmers and farmer associations, all with an initial focus on demonstration sites, from where positive results may, subsequently, be upscaled.

Final word

This book confirms the belief that inherent in smallholder farming communities are traditional knowledge systems which could usefully serve as a basis for developing resource management models (Brokensha, Warren, and Werner 1980; Scientific Advisory Group, 1994; Mammo, 1999). But because official policy tends to ignore it, this kind of knowledge is endangered. This underscores a need for a special policy attention to traditional knowledge, especially in regard to its documentation and nurturing for posterity.

REFERENCES

Brokensha, D. W., D. M. Warren, and O. Werner, eds, *Indigenous Knowledge Systems and Development*, Boston: University Press of America, 1980.

Mammo, T., *The Paradox of Africa's Poverty: The Role of Indigenous Knowledge, Traditional Practices and Local Institutions – The Case of Ethiopia*, Lawrenceville and Asmara: Red Sea Press, 1999.

Scientific Advisory Group, "Population [subsequently *People*], Land Management and Environmental Change (PLEC) – A short statement by the Scientific Advisory Group", *PLEC News and Views* 2:1, 1994.

Todaro, M. P., *Economic Development in the Third World*, New York: Longman, 1989.

Todaro, M. P., *Economic Development*, Reading, MA: Addison-Wesley, 2000.

Contributors

A. Sadik Abdulai (MSc), Lecturer, Department of Agricultural Mechanization and Irrigation Technology, University for Development Studies (UDS), Tamale, Ghana

Olivia Agbenyega (MPhil), Lecturer, Institute of Renewable Natural Resources (IRNR), Kwame Nkrumah University of Science and Technology (KNUST), Kumasi, Ghana

Diallo Amirou (PhD), Research Officer, Centre d' Etudes et de Récherche en Environnement, Université de Conakry, Conakry, Guinée

Charles Anane-Sakyi (MSc), Research Fellow, Soil Research Station, MangaBawku, Ghana

L. Asafo (MPhil), PhD Candidate, Department of Botany, University of Ghana, Legon, Ghana

Felix Asante (MPhil), Administrative and Research Officer, West African cluster of PLEC, c/o Department of Geography and Resource Development, University of Ghana, Legon, Ghana

William J. Asante (MSc), Lecturer, Department of Renewable Natural Resources, University for Development Studies (UDS), Tamale, Ghana

J. A. Bakang (PhD), Senior Lecturer, Faculty of Agriculture, Kwame Nkrumah University of Science and Technology (KNUST), Kumasi, Ghana

Essie T. Blay (PhD), Associate Professor, Department of Crop Science, University of Ghana, Legon, Ghana

Harold Brookfield (PhD), Emeritus Professor, Department of Anthropology, Division of Society and Environment, RSPAS, Australian National University, Canberra, ACT 0200, Australia

Stephen Nkansa Buabeng (MSc), Senior Research Fellow, Bureau of Integrated Rural Development (BIRD), Kwame Nkrumah University of Science and

Technology (KNUST), Kumasi, Ghana

D. Daouda (MSc), Research Assistant, Centre d' Etudes et de Récherche en Environnement, Université de Conakry, Conakry, Guinée

J. S. Dittoh (PhD), Associate Professor, Department of Agricultural Economics and Extension, and Pro-Vice-Chancellor, University for Development Studies (UDS), Tamale, Ghana

Lewis Enu-Kwesi (PhD), Associate Professor and Head, Department of Botany, University of Ghana, Legon, Ghana

B. Z. Gandaa (BSc), Senior Research Assistant, Department of Agricultural Mechanization and Irrigation Technology, University for Development Studies (UDS), Tamale, Ghana

Edwin A. Gyasi (PhD), Professor, Department of Geography and Resource Development, and Coordinating Leader, WAPLEC, University of Ghana, Legon, Ghana

J. Heloo, Country Director, Heifer Project International (HPI), Accra, Ghana

G. Kranjac-Berisavljevic (MSc), Senior Lecturer and Head, Department of Agricultural Mechanization and Irrigation Technology, University for Development Studies (UDS), Tamale, Ghana

Ebenezer Laing (PhD), Professor Emeritus, Department of Botany, University of Ghana, Legon, Ghana

Emmanuel Nartey, Farmer, Bormase, PO Box 39, Odumase-Krobo, Ghana

K. A. Nkyi (PhD), Lecturer, Institute of Renewable Natural Resources (IRNR), Kwame Nkrumah University of Science and Technology (KNUST), Kumasi, Ghana

W. Oduro (PhD), Senior Lecturer and Director, Institute of Renewable Natural Resources (IRNR), Kwame Nkrumah University of Science and Technology (KNUST), Kumasi, Ghana

Benjamin D. Ofori (MPhil), Research Fellow, Volta Basin Research Project, University of Ghana, Legon, Ghana

J. B. Ofori, Ministry of Food and Agriculture, Suhum, Ghana

Edward Ofori-Sarpong (PhD), Professor, Department of Geography and Resource Development, and Pro-Vice-Chancellor, University of Ghana, Legon, Ghana

Alfred A. Oteng-Yeboah (PhD), Associate Professor, Deputy Director General, Environment and Health Sector, Council for Scientific and Industrial Research (CSIR), Accra, Ghana

J. A. Poku (PhD), Director, Directorate of Crop Services, Ministry of Food and Agriculture, Accra, Ghana.

Charles Quansah (PhD), Associate Professor, Department of Crop Science, Kwame Nkrumah University of Science and Technology (KNUST), Kumasi, Ghana

P. B. Tanzubil (PhD), Officer-in-charge, Savanna Agricultural Research Station, Manga-Bawku, Ghana.

V. V. Vordzogbe (MPhil), Lecturer, Department of Botany, University of Ghana, Legon, Ghana

Index

Catalogue Request

Name: ______________________________

Address: ______________________________

____

Tel: ______________________________

Fax: ______________________________

E-mail: ______________________________

To receive a catalogue of UNU Press publications kindly photocopy this form and send or fax it back to us with your details. You can also e-mail us this information. Please put "Mailing List" in the subject line.

53-70, Jingumae 5-chome
Shibuya-ku, Tokyo 150-8925, Japan
Tel: +81-3-3499-2811 Fax: +81-3-3406-7345
E-mail: sales@hq.unu.edu http://www.unu.edu